Introduction to Mathematical Systems Theory

Christiaan Heij • André C. M. Ran •
Frederik van Schagen

Introduction to Mathematical Systems Theory

Discrete Time Linear Systems,
Control and Identification

Second Edition

 Birkhäuser

Christiaan Heij
Department of Econometrics
Erasmus University Rotterdam
Rotterdam, The Netherlands

André C. M. Ran
Department of Mathematics
Vrije Universiteit
Amsterdam, The Netherlands

Frederik van Schagen
Department of Mathematics
Vrije Universiteit
Amsterdam, The Netherlands

ISBN 978-3-030-59652-1   ISBN 978-3-030-59654-5   (eBook)
https://doi.org/10.1007/978-3-030-59654-5

Mathematics Subject Classification: 93-XX

1st edition: © Birkhäuser Basel 2007
2nd edition: © Springer Nature Switzerland AG 2021

This book is published under the imprint Birkhäuser, www.birkhauser-science.com, by the registered company Springer Nature Switzerland AG.
The registered company address is: Gewerbestrasse 11, 6330 Cham, Switzerland

Preface

This book has grown out of more than 15 years of lecturing an introductory course in system theory, control and identification for students in the areas of Business Mathematics and Computer Science, Econometrics and Mathematics at the Vrije Universiteit in Amsterdam. The interests and mathematical background of our students motivated our choice to focus on systems in discrete time only, because the topics can then be studied and understood without preliminary knowledge of (deterministic and stochastic) differential equations. This book does require some preliminary knowledge of calculus, linear algebra, probability and statistics, and some parts use the elementary results on Fourier series.

The book treats the standard topics of introductory courses in linear systems and control theory. Deterministic systems are discussed in the first five chapters, with the following main topics: realization theory, observability and controllability, stability and stabilization by feedback, and linear quadratic optimal control. Stochastic systems are treated in Chaps. 6 to 8, with main topics: realization, filtering and prediction (including the Kalman filter), and linear quadratic Gaussian optimal control. Chapters 9 and 10 discuss system identification and modelling from data, and Chap. 11 concludes with a brief overview of further topics.

Exercises form an essential ingredient of any successful course in this area. The exercises are not printed in the book and are instead incorporated on the accompanying CD-ROM. The exercises are of two types: that is, theory exercises to train mathematical skills in system theory and practical exercises applying system and control methods to data sets that are also included on the CD-ROM. Many exercises require the use of Matlab or a similar software package.

We did benefit greatly from the comments of many colleagues who, over the years, participated in teaching from this book. In particular, we like to mention the contributions of (in alphabetical order) Sanne ter Horst, Rien Kaashoek, Derk Pik, Jan H. van Schuppen and Alistair Vardy. We thank them for their comments, which have improved the text considerably. In addition, many students helped us in improving the text by asking questions and pointing out misprints.

Preface to the Second Edition

The second edition of the book is a slightly altered version of the first edition. We corrected many misprints and mis-statements that were present in the first edition. In addition, we made many small changes to the text to improve readability.

There are several more extensive changes as well. Chapter 3 has been expanded to incorporate two sections dealing with the subspace identification algorithm, providing a construction of a minimal realization directly from sequences of inputs and outputs. In Sect. 6.5, material was added to make the discussion of spectra of several types of processes more complete. Chapter 11 has been expanded with several parts dealing with modern developments. In connection with this, the list of references has been expanded considerably. We hope this will be useful by providing pointers for further study.

Changing technology has changed the way in which the exercises are presented. At the time of writing of the first edition, a CD-ROM was state of the art; however, present-day laptops are not equipped with a CD drive anymore. For that reason, we have decided to make the exercises available as a separate file on the SpringerLink's book website.

Rotterdam, The Netherlands   Christiaan Heij
Amsterdam, The Netherlands   André C. M. Ran
Amsterdam, The Netherlands   Frederik van Schagen

Electronic Supplementary Material The online version of this book (https://doi.org/10.1007/978-3-030-59654-5_11) contains supplementary material, which is available to authorized users.

Contents

Dynamical Systems

1

1.1 Introduction

Many phenomena investigated in such diverse areas as physics, biology, engineering, and economics show a dynamical evolution over time. Examples are thermodynamics and electromagnetism in physics, chemical processes and adaptation in biology, control systems in engineering, and decision making in macro economics, finance, and business economics. The main questions analysed in this book are the following.

- What type of mathematical models can be used to study such dynamical processes?
- Once a model class is selected and we know the parameters in the model, how can we achieve specific objectives such as stability, uncertainty reduction and optimal decision making?
- If we do not know the parameters in the model exactly, how can we estimate them from available data and how reliable is the obtained model?

The first question is the topic of Chaps. 2, 3 and 6, the second one of Chaps. 4, 5, 7, 8 and 9, and the third one of Chaps. 9 and 10. The answers to these questions will in general depend on accidental particularities of the problem at hand. However, there are important common characteristics of these problems which can be expressed in terms of mathematical models. We first give some examples to illustrate the main ideas in modelling, estimation, forecasting and control.

Example 1.1.1 Suppose that for a certain good the market functions as follows. The quantity currently produced will be supplied to the market in the next period. Supply and demand determine the market price. Let D denote the quantity demanded, S the quantity supplied, P the market price, $\hat{P}$ the anticipated price used by the suppliers in

C. Heij et al., *Introduction to Mathematical Systems Theory*,
https://doi.org/10.1007/978-3-030-59654-5_1

their production decisions, and let t denote the time period. A simple market model is given by the equations

$$D(t) = \alpha_0 + \alpha_1 P(t) \tag{1.1}$$

$$S(t) = \beta_0 + \beta_1 \hat{P}(t) \tag{1.2}$$

$$\hat{P}(t) = f(P(s); s \leq t - 1) \tag{1.3}$$

$$S(t) = D(t) \tag{1.4}$$

Here (1.1) and (1.2) are (linearized) behavioural equations, in general with $\alpha_1 < 0$ and $\beta_1 > 0$. Equation (1.3) describes how the suppliers predict future prices, and Eq. (1.4) expresses the equilibrium condition of market clearing. In practice Eqs. (1.1) and (1.2) are of course only approximations, and the same holds true for the equilibrium condition (1.4). The precise form of the forecasting function (1.3) will in general also be unknown. Depending on the specification of (1.3), different dynamical systems result with corresponding different evolutions of prices and quantities. Adaptive forecasts can be described as follows, with $0 < \lambda < 1$,

$$\hat{P}(t) = \hat{P}(t - 1) + \lambda \{ P(t - 1) - \hat{P}(t - 1) \}. \tag{1.5}$$

Another specification is to let the price consist of an unobserved permanent component, denoted by X, that it is affected by stochastic disturbances, u and v,

$$P(t) = X(t) + u(t); \qquad X(t) = X(t - 1) + v(t). \tag{1.6}$$

The price forecast could be obtained by minimizing the expected quadratic forecast error $E(P(t) - \hat{P}(t))^2$, where EX denotes the expectation of the stochastic variable X. This is called a prediction or filtering problem.

If the form of (1.3) has been specified, then the dynamical evolution of price and quantity depends on the model parameters. In practice these parameters are in general unknown. Let us denote the model error in $D(t)$ by $\varepsilon_D(t)$, and the model error in $S(t)$ by $\varepsilon_S(t)$. Such model errors ε_D and ε_S arise for several reasons, for example neglected relevant variables and misspecification of the functional form and of the dynamics. Incorporating these model errors into the behavioural equations (1.1, 1.2) leads to the following model

$$D(t) = \alpha_0 + \alpha_1 P(t) + \varepsilon_D(t) \tag{1.7}$$

$$S(t) = \beta_0 + \beta_1 \hat{P}(t) + \varepsilon_S(t) \tag{1.8}$$

System identification is concerned with the estimation of the unknown model parameters from available data on quantity and prices.

Example 1.1.2 National governments are confronted with the task of designing a macro economic policy. A much simplified version of this problem is the following model of the business cycle. Let C denote consumption expenditures, Y national income, I investments and G government expenditure. We assume that consumption depends on the income of the last period and that investments are based on the so-called accelerator principle. This gives the model

$$C(t) = \alpha + \beta Y(t - 1) \tag{1.9}$$

$$I(t) = \gamma + \delta\{C(t) - C(t - 1)\} \tag{1.10}$$

$$Y(t) = C(t) + I(t) + G(t) \tag{1.11}$$

Here (1.11) is a definitional equality, and Eqs. (1.9) and (1.10) are behavioural equations. From the viewpoint of macro economic policy this leaves the variable G as policy or control variable. In econometrics this is called an exogenous variable, in systems theory an input variable. The other variables Y, C and I are the policy targets. In econometrics these are called endogenous variables, in systems theory output variables.

The government could be interested in regulating income, consumption and investments. A possible objective is to keep these macro economic variables as close as possible to pre-assigned target trajectories, denoted by Y^*, C^* and I^*. If N denotes the planning horizon, then deviations from these objectives could, for instance, be measured by the cost function

$$J = \sum_{t=1}^{N} [g_1\{Y(t) - Y^*(t)\}^2 + g_2\{C(t) - C^*(t)\}^2 + g_3\{I(t) - I^*(t)\}^2]. \tag{1.12}$$

Here the coefficients g_i, $i = 1, 2, 3$, reflect the relative importance of the objectives. This is an example of a dynamic optimization problem, known as the linear quadratic control problem.

We should mention that in modern economics control theory plays a role mainly in the following two areas. First, in micro economic theory to model the behaviour of individual economic agents. Second, in business applications, for example in production planning and financial decision making. Macro economic models and control theory play only a minor role in government decisions. Originally such an approach was inspired by the wish to smooth business cycles. However, macro economic policy depends on many factors that are not easily captured in a model.

Example 1.1.3 Consider a firm producing a single good. The production is organized in planning periods of 3 months. At the beginning of each period the production quantity is determined. In order to meet random fluctuations in demand the firm also holds inventories of the good. Let D denote demand, Q the quantity produced, and X the inventory of the good. The inventory develops according to

$$X(t+1) = X(t) + Q(t) - D(t) \tag{1.13}$$

We assume that a negative inventory corresponds to excess demand that will be satisfied by the production in the next period. Let $f(X)$ denote the cost of holding inventory (if $X > 0$) and the cost of delayed demand (if $X < 0$). Further let α denote the production cost per unit, and let N be the planning horizon. As the demand shows random variations, the firm could minimize the expected total cost

$$J = E[\sum_{t=1}^{N} \{\alpha Q(t) + f(X(t+1))\}] \tag{1.14}$$

This is a stochastic optimal control problem. Here Q is the input or control variable. The optimal production plan depends on the cost function f and on the demand process D.

If the demand $D(t)$ contains trends and seasonal patterns, these may be incorporated in a model for the demand. In such a so-called structural model the demand is decomposed in a trend term T with varying slope B, a seasonal term S consisting of a yearly recurring pattern, and a random component ε. We then have equations for the trend, for the varying slope of the trend, and for the seasonal term S. The demand model is not driven by control variables, but by noise terms in each of the equations, representing unanticipated random shocks. The auxiliary variables $T(t)$, $B(t)$ and $S(t)$ are unobserved. A possible model of this kind is

$$D(t) = T(t) + S(t) + \varepsilon(t) \tag{1.15}$$

$$T(t) = T(t-1) + B(t-1) + \eta(t) \tag{1.16}$$

$$B(t) = B(t-1) + \xi(t) \tag{1.17}$$

$$S(t) = -S(t-1) - S(t-2) - S(t-3) + \omega(t) \tag{1.18}$$

Here ε, η, ξ and ω are random components. Note that (1.18) with $\omega(t) = 0$ gives a periodicity for $S(t)$. Indeed, starting with $S(0) = a, S(1) = b, S(2) = c$ leads to a repeating pattern in the sequence $S(t)$ of $a, b, c, -a - b - c$, as one readily checks.

1.2 Systems and Laws

In this section we shall consider several types of models that can be used to describe dynamical systems. A dynamical system is characterized by a collection of system variables that evolve in mutual dependence. Denoting the time axis by T and the outcome space for the system variables at each time instant by W, we formalize a (deterministic) dynamical system as follows.

Definition 1.2.1 A *dynamical system* consists of a set of allowable trajectories of the system variables, i.e., it is characterized by its behaviour $\mathcal{B} \subset \{w : T \to W\}$.

This gives a deterministic description, as trajectories in $\mathcal{B}$ are possible within the system and the other trajectories are excluded. In this book we only consider systems that evolve in discrete time ($T = \mathbb{Z}$). We will not discuss continuous time systems (with $T = \mathbb{R}$), as the theory of linear systems largely coincides for both cases. In economics one uses mostly discrete time models, because the available data are often observed in discrete time. The mathematical models for discrete time systems involve difference equations, whereas models for continuous time systems involve differential equations.

In many applications a system is considered as a part of reality which interacts with its environment. In this case the variables are divided into inputs, consisting of the variables from the environment that influence the system, and outputs, the variables that describe the effects. A system is then seen as a mechanism producing the outputs (endogenous variables) from the inputs (exogenous variables).

Definition 1.2.2 An *input-output system* consists of a set of input trajectories $\{u : T \to U\}$ and output trajectories $\{y : T \to Y\}$ related by a mapping F. The system behaviour is given by $\mathcal{B} = \{(u, y) : T \to U \times Y; y = F(u)\}$.

This definition is somewhat limited, as it requires that the input uniquely determines the output. Sometimes an additional effect of so-called initial conditions is allowed.

For purposes of forecasting and control it is of particular interest to consider causal input-output systems. This means that the present output is completely determined by the past evolution of the inputs (including the present input).

Definition 1.2.3 A *causal input-output system* is a system for which the input-output mapping F has the property that $y(t) = F_t(u(s); s \leq t)$ for certain mappings $F_t, t \in T$.

For identification and control of dynamical systems it is of crucial importance that the system is expressed in a convenient way. A given system can be represented in many alternative ways. Which representation is preferred will depend on the model objectives.

A system can often be described by a set of equations, the system laws, each of which describes a relationship between the variables. Of special interest are (vector) difference equations of the form

$$G(t, w(t), w(t-1), \ldots, w(t-L)) = 0, \tag{1.19}$$

where $w(t)$ is a vector in $\mathbb{R}^q$ for all $t \in \mathbb{Z}$, and where G is a map from $\mathbb{Z} \times (\mathbb{R}^q)^{L+1}$ to $\mathbb{R}^p$ for some p. The parameter L specifies the order of the equation. If the system laws are invariant over time then they can be expressed in the form

$$G(w(t), w(t-1), \ldots, w(t-L)) = 0 \tag{1.20}$$

Equations of this type (and their continuous time counterparts, being differential equations) are at the heart of most dynamical models in physics, biology, engineering, economics, and other sciences. The system is called linear if the function G is linear in its arguments. Linear difference equations are very useful in practice for the following reasons.

- In many cases nonlinear systems can be approximated rather well by linear ones.
- In most applications the precise nature of possible nonlinearities is unclear, but methods based on nonlinear models are often sensitive with respect to the chosen type of nonlinearity.
- Practical identification and control methods are mostly developed for linear systems.

Nonlinear systems can be approximated by linearization. Let w_0 be a solution of interest and let $A_k(t)$ be the matrix of first derivatives of G in (1.20) with respect to the k-th position, $k = 0, \ldots, L$, and evaluated at w_0 and time t. If w has q variables and G consists of p equations then these matrices have size $p \times q$. Locally around the solution w_0, other system trajectories w satisfying (1.20) are approximately described by the first order Taylor expansion in terms of $w_\triangle := w - w_0$, that is

$$A_0(t)w_\triangle(t) + A_1(t)w_\triangle(t-1) + \ldots + A_L(t)w_\triangle(t-L) \approx 0 \tag{1.21}$$

This is a linear difference equation with time-varying parameters. If w_0 is constant over time, for instance identically zero, then the parameters are constant, and locally around zero the system is described by the approximate law

$$A_0 w(t) + A_1 w(t-1) + \ldots + A_L w(t-L) = 0 \tag{1.22}$$

Systems of the form (1.22) are called linear, time invariant and finite dimensional. These systems are of fundamental importance in identification and control. Linearity means that the behaviour defined by the solution set of equation (1.22) is a linear space. Time invariance means that solutions shifted in time remain within the system. The property of finite dimensionality has to do with state space models, as discussed in the next section.

Because of their relative simplicity and because good theory for solving systems of the type (1.22) exists, these models are very useful.

We illustrate the concept of linearization with a well-known example from physics, for once using a continuous time model. The mathematical model for a swinging pendulum with no friction is given by the differential equation $x(t)'' = c \cdot \sin(x(t))$. Here $x(t)$ denotes the angle of the swinging pendulum with the downward vertical measured in radians and c is a constant depending on the length of the pendulum. The pendulum is at rest when $x(t) \equiv 0$; this is called an equilibrium solution. Linearization around this equilibrium solution is obtained by observing that for small values of $x(t)$ one has $\sin(x(t)) \approx x(t)$, whereby the differential equation for the linearized model becomes a linear equation: $x(t)'' = c \cdot x(t)$.

1.3   State Representations

In the foregoing section the system structure was made explicit by means of functional relationships between the system variables. In order to derive and use these relationships it is often helpful to introduce auxiliary variables. An auxiliary variable of particular importance is the state, which summarizes all the past information that is relevant for the future evolution of the system. State space models have the following form, where A, B, C, D are matrices of appropriate dimensions.

$$x(t + 1) = Ax(t) + Bv(t) \tag{1.23}$$

$$w(t) = Cx(t) + Dv(t) \tag{1.24}$$

Here v is an auxiliary variable which drives the evolution of the state x via the first order equation (1.23), with resulting observed system trajectory w described by the static equation (1.24). To describe the evolution of w for future times, $t \geq t_0$, it suffices to know $x(t_0)$ and the future driving forces $v(t), t \geq t_0$. So $x(t_0)$ summarizes the past information that is relevant for the future. For this reason x is called a state variable.

Definition 1.3.1 A *state space representation of a system* with variables w is a representation of the form (1.23), (1.24), so that the behaviour is given by

$$\mathcal{B} = \{w : T \rightarrow W; \text{ there exist } (x, v) \text{ such that (1.23) and (1.24) are satisfied }\}.$$

The practical usefulness of state space models lies in their simple first order dynamical structure. It can be shown that it is precisely the class of systems described by (1.22) which can be represented in state space form. As the state vector contains a finite number of elements, these systems are called finite dimensional.

Stochastic state space systems are of the form (1.23), (1.24) with w an observed stochastic process and with v an auxiliary white noise process. In this case the process x has the Markov property and acts as a sufficient statistic in prediction and control. The class of stochastic processes which can be represented in this way corresponds to the widely used class of so-called autoregressive moving average processes.

For causal input-output systems satisfying Eq. (1.22) we obtain, by splitting the system variables w into inputs u and outputs y, a representation of the form

$$P_0 y(t) + P_1 y(t-1) + \ldots + P_L y(t-L) = Q_0 u(t) + Q_1 u(t-1) + \ldots + Q_L u(t-L) \qquad (1.25)$$

Causality is guaranteed if P_0 is invertible. A state space representation of such a system can be obtained with the input variables u as the auxiliary variables v. As we shall see later, this leads to the following so-called *input-state-output representation*

$$x(t+1) = Ax(t) + Bu(t), \qquad (1.26)$$

$$y(t) = Cx(t) + Du(t). \qquad (1.27)$$

This model forms the corner stone in linear control theory. In the next chapter we shall show the equivalence of the representations (1.25) and (1.26, 1.27) for appropriately chosen matrices A, B, C and D.

If we neglect the influence of initial conditions and consider the effect of the inputs on the outputs, then this relationship can be described by a convolution, i.e., under certain boundedness conditions we get

$$y(t) = \sum_{k=0}^{\infty} G_k u(t-k). \qquad (1.28)$$

Convolution systems are sometimes more easily analysed in the so-called frequency domain, where the system trajectories are decomposed into cyclical components. Stochastic processes allow a similar decomposition into cyclical components.

1.4   Illustration

To give a simple illustration of the concepts and models discussed in this chapter we consider the business cycle model of Example 1.1.2. Suppose that the purpose of this model is to describe the macro economic business cycle in Y, C and I, and the possible effects of government spending G.

This system, in the sense of Definition 1.1, is characterized by the solution set of the three Eqs. (1.9)–(1.11). The equilibrium values, for a fixed level of government expenditure $\overline{G}$, are obtained by solving the equations for fixed levels of all the variables, so that

$$\overline{C} = \frac{\alpha + \beta\gamma + \beta\overline{G}}{1 - \beta}, \quad \overline{I} = \gamma, \quad \overline{Y} = \frac{\alpha + \gamma + \overline{G}}{1 - \beta}. \tag{1.29}$$

Defining the deviations from equilibrium by $c(t) = C(t) - \overline{C}, i(t) = I(t) - \overline{I}, y(t) = Y(t) - \overline{Y}$ and $g(t) = G(t) - \overline{G}$, we obtain the linear model

$$c(t) = \beta y(t - 1), \tag{1.30}$$

$$i(t) = \delta\{c(t) - c(t - 1)\}, \tag{1.31}$$

$$y(t) = c(t) + i(t) + g(t). \tag{1.32}$$

Let us consider what happens if we take consumption as an input variable in the sense of Definition 1.2.2. Specification of $c(t)$ for all $t \in \mathbb{Z}$ gives unique corresponding values of $y(t), i(t)$ and $g(t)$, for all $t \in \mathbb{Z}$. However, this is not a causal input-output system as $\{c(t); t \leq t_0\}$ determines $\{i(t); t \leq t_0\}, \{y(t); t \leq t_0 - 1\}$ and $\{y(t) - g(t); t \leq t_0\}$. Hence it is not possible to determine $y(t_0)$ and $g(t_0)$.

A causal input-output system can be obtained, for instance, by taking government expenditure as input. This can be seen from the relationship

$$y(t) = \beta(1 + \delta)y(t - 1) - \beta\delta y(t - 2) + g(t), \tag{1.33}$$

which one readily checks. This shows that income, consumption and investment are determined in a causal way by government expenditure for two given initial values of income. When the model is used to explain the business cycle in income and the effect of government policy, then Eq. (1.33) is the core of the model after the auxiliary variables c and i have been eliminated. This so-called "final form equation" contains all dynamical information on the income process. Note that Eq. (1.33) would probably not easily have been specified on theoretical grounds, but it is motivated by the auxiliary behavioural relations (1.30) and (1.31).

An input-output description of the form (1.25) is easily obtained:

$$\begin{pmatrix} 1 & 0 & 0 \\ -\delta & 1 & 0 \\ -1 & -1 & 1 \end{pmatrix} \begin{pmatrix} c(t) \\ i(t) \\ y(t) \end{pmatrix} + \begin{pmatrix} 0 & 0 & -\beta \\ \delta & 0 & 0 \\ 0 & 0 & 0 \end{pmatrix} \begin{pmatrix} c(t - 1) \\ i(t - 1) \\ y(t - 1) \end{pmatrix} = \begin{pmatrix} 0 \\ 0 \\ 1 \end{pmatrix} g(t). \tag{1.34}$$

An input-state-output representation is obtained from this by taking as past information the state variable $x(t)$ with components $c(t-1)$, $i(t-1)$, and $y(t-1)$, so that

$$x(t+1) = \begin{pmatrix} 0 & 0 & \beta \\ -\delta & 0 & \beta\delta \\ -\delta & 0 & \beta(1+\delta) \end{pmatrix} x(t) + \begin{pmatrix} 0 \\ 0 \\ 1 \end{pmatrix} g(t), \tag{1.35}$$

$$\begin{pmatrix} c(t) \\ i(t) \\ y(t) \end{pmatrix} = \begin{pmatrix} 0 & 0 & \beta \\ -\delta & 0 & \beta\delta \\ -\delta & 0 & \beta(1+\delta) \end{pmatrix} x(t) + \begin{pmatrix} 0 \\ 0 \\ 1 \end{pmatrix} g(t). \tag{1.36}$$

This state representation is not minimal, in the sense that there exist representations with fewer state variables. For example, by taking $z(t) = (y(t-1), y(t-2))^T$ we get the model

$$z(t+1) = \begin{pmatrix} \beta(1+\delta) & -\beta\delta \\ 1 & 0 \end{pmatrix} z(t) + \begin{pmatrix} 1 \\ 0 \end{pmatrix} g(t), \tag{1.37}$$

$$\begin{pmatrix} c(t) \\ i(t) \\ y(t) \end{pmatrix} = \begin{pmatrix} \beta & 0 \\ \beta\delta & -\beta\delta \\ \beta(1+\delta) & -\beta\delta \end{pmatrix} z(t) + \begin{pmatrix} 0 \\ 0 \\ 1 \end{pmatrix} g(t). \tag{1.38}$$

It can be shown that this is a minimal state representation. We shall return to this example in the next chapter.

Input-Output Systems

2

In this chapter we consider input-output systems. Such systems can be described in the time domain, in terms of the impulse response, and in the frequency domain, by the transfer function. For rational transfer functions the system can be represented by a finite dimensional state space model.

2.1 Inputs and Outputs in the Time Domain

Dynamical systems are characterized by a collection of variables and their interrelationships over time. As stated before, we shall only consider discrete time systems. For input-output systems there are two types of variables, namely inputs, which one may choose freely and outputs, which are determined by the choice of the inputs. Such systems may be schematically described by the following figure:

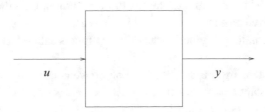

Here $u(t) = \left(u_1(t) \ldots u_m(t) \right)^T$ is the vector with the m input variables at time t and $y(t) = \left(y_1(t) \ldots y_p(t) \right)^T$ is the vector with the p output variables. The box is called the plant or the process and stands for the way the outputs depend on the inputs. If $m = p = 1$, then the system is called a single input, single output (SISO) system, and all other cases

C. Heij et al., *Introduction to Mathematical Systems Theory*,
https://doi.org/10.1007/978-3-030-59654-5_2

are referred to as multi-variable systems. If the input variables are meant to control the system they are called control or command variables. It may also be that some of the input variables are not in our hands. Such inputs may result from outside disturbances.

If the input is given beforehand, this is an open loop system. One may wish to choose the input in such a way that a prescribed goal is reached. This is called an optimal control problem. If the input is regulated by information on the past evolution of the system then this is a closed loop system. This may be depicted as follows.

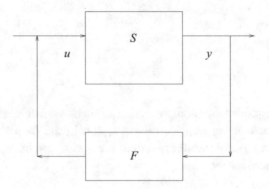

Example 2.1.1 We consider the business cycle model of Sect. 1.4. The system equations (1.30)–(1.32) lead to the following relation between national income $y(t)$ and government spending $g(t)$

$$y(t) - \beta(1 + \delta)y(t - 1) + \beta\delta y(t - 2) = g(t). \tag{2.1}$$

Here we may think of y as the output and of g as the input of the system. If the government spends without regarding the income this is an open loop system. If spending is seen as an instrument to steer the income in a desired path of evolution, then the value of $g(t)$ may be based on the realized past incomes $y(t - 1)$, $y(t - 2)$, ... This is a closed loop system. The use of past information to generate the current input is called feedback.

We denote a system by the symbol Σ. For convenience we assume that Σ starts operating at time instant $t = 0$. For time $t < 0$ the values of all variables are supposed to be zero. Since the system operates in discrete time, the *input trajectories* are sequences $u = (u(0), u(1), \ldots)$ of which the elements are vectors in $\mathbb{R}^m$. The *output trajectories* are sequences $y = (y(0), y(1), \ldots)$ with elements in $\mathbb{R}^p$. The map which assigns to a given input trajectory u the corresponding output trajectory y is called the *input-output map* and is denoted by G_Σ. So the output $y(t)$ is given by $y(t) = (G_\Sigma u)(t)$.

The system is called *causal* if the output does not depend on future inputs and is totally dependent on past and present inputs, that is, if for any two input trajectories u and v the following implications holds for each t_0:

$$\{u(t) = v(t), \quad \forall t \le t_0\} \Rightarrow \{(G_\Sigma u)(t) = (G_\Sigma v)(t), \quad \forall t \le t_0\}.$$

The system Σ is said to be *linear* if the input-output map is a linear transformation, i.e.,

$$G_\Sigma(u + v) = G_\Sigma u + G_\Sigma v, \qquad G_\Sigma(\lambda u) = \lambda G_\Sigma u \quad (\lambda \in \mathbb{R}).$$

If Σ is linear and causal, then the output $y(t)$ at time k depends linearly on the inputs $u(0), \ldots, u(t)$. It follows that the input-output map can be written in the form

$$(G_\Sigma u)(t) = \sum_{k=0}^{t} G(t, k)u(k), \qquad t = 0, 1, 2, \ldots. \tag{2.2}$$

where $G(t, k)$, $0 \le k \le t$ are $p \times m$ matrices.

The system Σ is said to be *time-invariant* if the input-output behaviour does not depend on time. That is, if an input trajectory u is applied k time instants later, then the resulting output trajectory is the original output trajectory but now also starting k time instants later. To make this more precise, we consider the shift operator S on trajectories, defined by

$$S(x(0), x(1), x(2), \ldots) = (0, x(0), x(1), \ldots).$$

The condition that the system Σ is time-invariant is equivalent to

$$G_\Sigma S = SG_\Sigma. \tag{2.3}$$

(Strictly speaking, we abuse notation here. The letter S on the left hand side of the equality is the shift on trajectories of input vectors, while on the right-hand side of the equality it denotes the shift on trajectories of output vectors.)

Proposition 2.1.1 *Let Σ be a causal linear time-invariant system. Then the input-output map of Σ has the following form*

$$(G_\Sigma u)(t) = \sum_{k=0}^{t} G(t - k, 0)u(k), \qquad t \ge 0. \tag{2.4}$$

Proof   Let e be an *impulse* applied at time $t = 0$, so $e = (u(0), 0, 0, \ldots)$, with $u(0)$ an arbitrary vector in $\mathbb{R}^m$. Using (2.2) it follows that

$$(S^k G_\Sigma e)(t) = G(t - k, 0)u(0), \quad t \geq k, \quad \text{while} \quad (G_\Sigma S^k e)(t) = G(t, k)u(0), \quad t \geq k.$$

Thus, using (2.3) repeatedly, we see that $G(t, k) = G(t - k, 0)$.   □

From now on we shall denote $G(t, 0)$ by $G(t)$. For a causal linear time-invariant system Σ the sequence of matrices $(G(0), G(1), G(2), \ldots)$ is called the *impulse response of the system*. This terminology is motivated by the output response to an impulse input:

$$G_\Sigma(u(0), 0, 0, \ldots) = (G(0)u(0), G(1)u(0), G(2)u(0), \ldots). \tag{2.5}$$

Example 2.1.2   Consider again the relation (2.1) between government spending $g(t)$ and national income $y(t)$. It is easily checked that this is a causal, linear, time invariant system if we take g as input and y as output. We suppose that the economy starts at equilibrium, so that $y(t) = g(t) = 0$ for $t < 0$, see Sect. 1.4. One can deduce from (2.1) that the impulse response of this system is given by $G(0) = 1$, $G(1) = \beta(1 + \delta)$, whereas for $k \geq 2$ the impulse response matrices satisfy the second order difference equation

$$G(k) = \beta(1 + \delta)G(k - 1) - \beta\delta G(k - 2), \qquad k \geq 2.$$

(The verification of this is left to the reader as an exercise.) Hence they can be computed recursively.

2.2   Frequency Domain and Transfer Functions

Let $u = (u(0), u(1), \ldots)$ be a sequence with elements $u(t) \in \mathbb{R}^m$. By definition the *z-transform* of u is the formal power series in z^{-1} given by

$$\widehat{u}(z) := u(0) + \frac{1}{z}u(1) + \frac{1}{z^2}u(2) + \cdots. \tag{2.6}$$

This is a formal power series in the sense that it is not required that the series in the right-hand side of (2.6) converges. In a similar way, if $G = (G(0), G(1), \ldots)$ is a sequence of $p \times m$ matrices then the *z*-transform of G is the formal power series in z^{-1} given by

$$\widehat{G}(z) = G(0) + \frac{1}{z}G(1) + \frac{1}{z^2}G(2) + \cdots. \tag{2.7}$$

For $\widehat{u}(z)$ and $\widehat{G}(z)$ as above the product $\widehat{G}(z)\widehat{u}(z)$ is defined as the formal power series which one obtains by carrying out the product formally,

$$\widehat{G}(z)\widehat{u}(z) = \sum_{k=0}^{\infty} \left(\frac{1}{z}\right)^k (G(k)u(0) + G(k-1)u(1) + \cdots + G(0)u(k)). \qquad (2.8)$$

To discuss convergence of the sequence in (2.8) we first introduce the following notion. A sequence $x = (x(0), x(1), \ldots)$ of vectors or matrices is *exponentially bounded* if there exist positive constants M and α such that

$$\|x(t)\| \le M\alpha^t, \qquad t = 0, 1, 2, \ldots. \qquad (2.9)$$

The norm in (2.9) is the Euclidean norm for vectors, that is, if the vector is given by $x(t) = \left(x_1(t) \cdots x_n(t)\right)^T$ then $\|x(t)\|^2 = \sum_{j=1}^{n} |x_j(t)|^2$. The norm in (2.9) is the induced matrix norm if $x(t)$ are matrices, i.e., if $x(t)$ is an $p \times q$ matrix then

$$\|x(t)\| = \sup_{\|w\|=1, \; w \in \mathbb{R}^q} \|x(t)w\|.$$

An important property of the induced matrix norm is the inequality

$$\|Ax\| \le \|A\| \cdot \|x\|$$

for any matrix A and any vector x (provided, of course, the product Ax makes sense), and

$$\|AB\| \le \|A\| \cdot \|B\|$$

for any pair of matrices A and B for which the product AB can be formed. Under this condition, the series $x(0) + \frac{1}{z}x(1) + \frac{1}{z^2}x(2) + \ldots$ is convergent for $z \in \mathbb{C}$ with $|z| > \alpha$, and in that case $\widehat{x}(z)$ is a well-defined function on $|z| > \alpha$.

If the impulse response sequence $G(j), \; j = 0, 1, \ldots$ is exponentially bounded, then

$$\widehat{G}(z) = \sum_{k=0}^{\infty} \frac{1}{z^k} G(k),$$

is called the *transfer function* of the system.

Proposition 2.2.1 *Let Σ be a causal linear time invariant system with exponentially bounded impulse response sequence $(G(0), G(1), \ldots)$. If $u = (u(0), u(1), \ldots)$ is an exponentially bounded input trajectory, then the output trajectory $y = G_\Sigma u$ is also* *exponentially bounded, and for $|z|$ sufficiently large we have*

$$\widehat{y}(z) = \widehat{G}(z)\widehat{u}(z). \tag{2.10}$$

Proof Suppose that $\|u(t)\| \leq M_1\alpha_1^t$ and $\|G(t)\| \leq M_2\alpha_2^t$. We may always take $\alpha_2 > \alpha_1$. From Proposition 2.1.1 it follows that the sequence of outputs $y(t)$ is given by $y(t) = \sum_{j=0}^{t} G(t-j)u(j)$ (cf. (2.4)), and therefore

$$\|y(t)\| \leq \sum_{k=0}^{t} M_2\alpha_2^{t-k} M_1\alpha_1^k = M_2 M_1 \frac{\alpha_2^{t+1} - \alpha_1^{t+1}}{\alpha_2 - \alpha_1} \leq M_1 M_2 \frac{\alpha_2}{\alpha_2 - \alpha_1}\alpha_2^t = M_3\alpha_2^t,$$

where $M_3 = M_1 M_2 \frac{\alpha_2}{\alpha_2 - \alpha_1}$. Thus the sequence of outputs is exponentially bounded. Since $\sum y(t)z^{-t}$, $\sum u(t)z^{-t}$ and $\sum G(t)z^{-t}$ converge absolutely for $|z|$ big enough, we conclude from a well-known result in analysis (see, e.g., [65, Theorem 3.50]) that (2.10) holds for $z \in \mathbb{C}$, with $|z|$ sufficiently large.   □

When we apply the z-transform to a trajectory we say that we pass from the *time domain* to the so-called *frequency domain*. In the time domain the action of the system is given by the input-output map which is the somewhat complicated convolution (2.4). In the frequency domain the action of the system is given by a straightforward multiplication, see (2.10). To justify the use of the term "frequency domain" consider the following example.

Example 2.2.1 The transfer function describes the way in which the frequencies in the inputs are transferred to the outputs. We illustrate this for a SISO system with impulse-response sequence satisfying $|G(k)| \leq M\alpha^k$ for some $\alpha < 1$. In particular $\sum_{k=0}^{\infty} |G(k)| < \infty$. Assume we have a summable input trajectory, that is, $\sum_{k=0}^{\infty} |u(k)| < \infty$. Then $u(t)$ has a well-defined Fourier transform $\widehat{u}(e^{i\omega}) = \sum_{t=0}^{\infty} u(t)e^{-i\omega t}$. The same holds for $G(t)$ and $y(t)$, and, moreover, we have $\widehat{y}(e^{i\omega}) = \widehat{G}(e^{i\omega})\widehat{u}(e^{i\omega})$. In particular, the absolute value of $\widehat{G}(e^{i\omega})$ is called the *gain*, and this number shows the amplification of the frequency $\frac{2\pi}{\omega}$ by the system.

2.3 State Space Models

The input-output map T of a system has a *state space representation* if the action $Tu = y$ can be described by a system of equations of the following type:

$$\begin{cases} x(t+1) = Ax(t) + Bu(t), \quad t = 0, 1, 2, \ldots, \\ y(t) = Cx(t) + Du(t), \\ x(0) = 0. \end{cases} \tag{2.11}$$

As before, the system is assumed to be at rest for $t < 0$, so that $y(t) = 0$, $u(t) = 0$, and $x(t) = 0$ for $t < 0$. Further, A is a linear transformation acting on an n-dimensional Euclidean space $\mathbb{R}^n$, called the *state space*, and $B : \mathbb{R}^m \to \mathbb{R}^n$, $C : \mathbb{R}^n \to \mathbb{R}^p$, $D : \mathbb{R}^m \to \mathbb{R}^p$ are linear transformations. Choosing standard bases in Euclidean space, A, B, C and D correspond to matrices with real coefficients. A is called the *state transition matrix*, B the *input matrix*, C the *output matrix*, and D the *external* (or *feedthrough*) matrix.

Example 2.3.1 Consider the equation

$$y(t) - \beta(1 + \delta)y(t - 1) + \beta\delta y(t - 2) = g(t) \tag{2.12}$$

of Example 2.1.1 for the relation between government spending $g(t)$ and national income $y(t)$. If we assume $y(t_0 - 1)$ and $y(t_0 - 2)$ to be known, then for $t \geq t_0$ the values of $y(t)$ are uniquely determined by $g(t)$ for $t \geq t_0$. Thus the vector

$$x(t) = \begin{pmatrix} y(t - 1) \\ y(t - 2) \end{pmatrix}$$

describes the "state" of the economy at year t, and the input-output map corresponding to (2.12) is described by the following state space equations:

$$\begin{cases} x(t + 1) = \begin{pmatrix} \beta(1 + \delta) & -\beta\delta \\ 1 & 0 \end{pmatrix} x(t) + \begin{pmatrix} 1 \\ 0 \end{pmatrix} g(t), & t \geq t_0, \\ y(t) = \left(\beta(1 + \delta) \ -\beta\delta \right) x(t) + g(t). \end{cases} \tag{2.13}$$

It is left as an exercise to the reader to verify this.

If in (2.11), the state $x(t_0)$ is known, then for $t \geq t_0$ the state $x(t)$ is given by

$$x(t) = A^{t-t_0}x(t_0) + \sum_{j=t_0}^{t-1} A^{t-j-1}Bu(j), \qquad t > t_0. \tag{2.14}$$

Consequently, the output $y(t)$ is given by

$$y(t) = Du(t) + CA^{t-t_0}x(t_0) + \sum_{j=t_0}^{t-1} CA^{t-j-1}Bu(j).$$

The information about the past contained in the state $x(t_0)$ together with the input trajectory for $t \geq t_0$ allows us to determine the output trajectory for $t \geq t_0$. A *state variable* is a vector function $x(t)$ with the property that for each $t > t_0 \geq 0$ the output trajectory $(y(t_0), \ldots, y(t))$ is uniquely determined by the state $x(t_0)$ and the input trajectory $(u(t_0), \ldots, u(t))$. In particular, we need no information on the past inputs and outputs $u(s), y(s)$ for $s < t_0$. To achieve an effective reduction in the required past information, the initial state $x(t_0)$ and the input trajectory $(u(t_0), \ldots, u(t-1))$ should also uniquely determine the state $x(t)$. The variable $x(t)$ in (2.11) has this additional property, because of (2.14).

Theorem 2.3.1 *The state space model (2.11) describes the input-output map of a causal linear time-invariant system with impulse response*

$$
G(t) = \begin{cases} D & \text{for } t = 0, \\ C A^{t-1} B & \text{for } t > 0, \end{cases}
$$

and with transfer function equal to

$$
\widehat{G}(z) = D + C(z - A)^{-1} B. \tag{2.15}
$$

In formula (2.15) we use the notation $z - A$ to denote $z \cdot I - A$. This convention will be used frequently in the sequel.

Proof   The solution of the state equation $x(t+1) = Ax(t) + Bu(t), t \geq 0$, with $x(0) = 0$ is given by $x(t) = \sum_{k=0}^{t-1} A^{t-1-k} B u(k)$, for $t \geq 1$. So the input-output map T of the state space model (2.11) is given by $(Tu)(t) = \sum_{k=0}^{t-1} C A^{t-1-k} B u(k) + Du(t)$ (compare (2.2) where T is denoted by G_Σ).

It remains to prove the formula for the transfer function. The sequence $C A^{k-1} B, k = 1, 2, \ldots$ is exponentially bounded because $\|C A^{k-1} B\| \leq \|C\| \cdot \|B\| \cdot \|A\|^{k-1} = M \alpha^{k-1}$, where $M = \|C\| \cdot \|B\|$ and $\alpha = \|A\|$. It follows that (2.11) has a well-defined transfer function,

$$
\widehat{G}(z) = D + \sum_{k=1}^{\infty} \frac{1}{z^k} C A^{k-1} B. \tag{2.16}
$$

For $|z| > \|A\|$ we have $I = (z - A) \sum_{k=1}^{\infty} \frac{1}{z^k} A^{k-1}$, and hence $(z - A)^{-1} = \sum_{k=1}^{\infty} \frac{1}{z^k} A^{k-1}$, where the convergence of the series on the right-hand side is interpreted in the norm sense, i.e., we say that the series converges to a matrix E when $\lim_{n \to \infty} \|E - \sum_{k=1}^{n} \frac{1}{z^k} A^{k-1}\| = 0$. So for $|z| > \|A\|$ the series on the right-hand side of (2.16) converges to $C(z - A)^{-1} B$ which proves (2.15). $\square$

State space models can also be characterized in terms of transfer functions. A matrix valued function W is called a *rational matrix function* if its entries are quotients of polynomials. It is called *proper* if $\lim_{|z| \to \infty} W(z)$ exists.

Proposition 2.3.2 *The transfer function of the system represented by (2.11) is a rational matrix function that is proper.*

Proof By Cramer's rule it follows that $(z - A)^{-1}$ is a rational matrix function, and as $\lim_{|z| \to \infty} (z - A)^{-1} = 0$, it is also proper. □

Conversely, every system with a proper rational transfer function has a state space representation. this is what we shall show in the next theorem.

Theorem 2.3.3 *The input-output map of a causal linear time-invariant system Σ admits a state space representation if and only if Σ has a transfer function that is rational and proper.*

To prove this result we use the following lemma.

Lemma 2.3.4 *Consider the matrix polynomials $H(z) = \sum_{j=0}^{\ell-1} z^j H_j$ and $L(z) = z^\ell I + \sum_{j=0}^{\ell-1} z^j A_j$ of sizes $p \times m$ and $m \times m$, respectively. Let*

$$
A = \begin{pmatrix} 0 & I & & \\ & & \ddots & \\ & & & I \\ -A_0 & -A_1 & \dots & -A_{\ell-1} \end{pmatrix}, \quad B = \begin{pmatrix} 0 \\ \vdots \\ 0 \\ I \end{pmatrix}, \quad C = \begin{pmatrix} H_0 & H_1 & \dots & H_{\ell-1} \end{pmatrix}.
$$

Then $H(z)L(z)^{-1} = C(z - A)^{-1}B$ for z not an eigenvalue of A.

Proof First we assume that $H(z) \equiv I$, so that $C = (I, 0, 0, \dots, 0)$. Let $z \in \mathbb{C}$ and consider the equation

$$
L(z)x = y \tag{2.17}
$$

where $x, y \in \mathbb{C}^m$. Defining $x_1 = x$, $x_2 = zx_1, \dots, x_\ell = zx_{\ell-1}$, Eq. (2.17) can be rewritten as $-A_0 x_1 - A_1 x_2 - \dots - A_{\ell-1} x_\ell = zx_\ell - y$. Therefore

$$
A \begin{pmatrix} x_1 \\ \vdots \\ x_\ell \end{pmatrix} = z \begin{pmatrix} x_1 \\ \vdots \\ x_\ell \end{pmatrix} - By, \quad x = x_1 = C \begin{pmatrix} x_1 \\ \vdots \\ x_\ell \end{pmatrix}.
$$

If z is not an eigenvalue of A, then

$$
\begin{pmatrix} x_1 \\ \vdots \\ x_\ell \end{pmatrix} = (z - A)^{-1} B y,
$$

and hence $x = C(z - A)^{-1} B y$, which proves the lemma for the case $H(z) \equiv I$.

Next we consider the general case. Let

$$
\begin{pmatrix} C_1(z) \\ \vdots \\ C_\ell(z) \end{pmatrix} = (z - A)^{-1} B.
$$

From the previous result we know that $C_1(z) = L(z)^{-1}$. Since

$$
(z - A) \begin{pmatrix} C_1(z) \\ \vdots \\ C_\ell(z) \end{pmatrix} = \begin{pmatrix} 0 \\ \vdots \\ 0 \\ I \end{pmatrix},
$$

the special form of A implies that $C_j(z) = z^{j-1} C_1(z)$, $j = 1, \ldots, \ell$. Hence

$$
C(z - A)^{-1} B = \begin{pmatrix} H_0 & \ldots & H_{\ell-1} \end{pmatrix} \begin{pmatrix} C_1(z) \\ z C_1(z) \\ \vdots \\ z^{\ell-1} C_1(z) \end{pmatrix} = H(z) C_1(z) = H(z) L(z)^{-1},
$$

which proves the Lemma.   □

Proof (Proof of Theorem 2.3.3) Proposition 2.3.2 already shows that the transfer function of a state space model is rational.

To prove the converse, let Σ be a causal linear time-invariant system with rational transfer function $\widehat{G}(z)$ that is proper. We have to prove that $y(t) = (G_\Sigma u)(t) = \sum_{k=0}^{t} G(t-k) u(k)$, $t = 0, 1, 2, \ldots$, admits a state space representation. As $\widehat{G}(z)$ is rational and proper, each entry is rational and proper as well. So $\widehat{G}(z) \to G(0)$ for $|z| \to \infty$, and hence we can write $\widehat{G}(z) = G(0) + K(z)$, where

$$
\lim_{|z| \to \infty} K(z) = 0. \tag{2.18}
$$

The (i, j)-th entry $k_{ij}(z)$ of $K(z)$ is a quotient of two polynomials, $k_{ij}(z) = p_{ij}(z)/q_{ij}(z)$, where we take $q_{ij}(z)$ to be monic, that is, with leading coefficient equal to 1. Condition (2.18) implies that the degree of p_{ij} is strictly less than the degree of q_{ij}. Let $r(z)$ be the product of all $q_{ij}(z)$, that is, $r(z) = \Pi_{ij}q_{ij}(z)$ and $H(z) = r(z)K(z)$, then $H(z)$ is a matrix polynomial, $H(z) = H_0 + zH_1 + \cdots + z^{\ell-1}H_{\ell-1}$, and from the property of the degrees of p_{ij} and q_{ij} it follows that ℓ is not larger than the degree of $r(z)$. Define $L(z) = r(z)I$, where I is the $m \times m$ identity matrix, then $L(z)$ is a monic matrix polynomial (i.e., its leading coefficient is I) and $K(z) = H(z)L(z)^{-1}$. Thus, by Lemma 2.3.4 we can find matrices A, B and C so that $K(z) = C(z - A)^{-1}B$, whenever z is not an eigenvalue of A. This shows that $G(t) = CA^{t-1}B$ for $t \geq 1$, so that G_Σ has a state space representation (with $D = G(0)$). $\qquad \square$

The choice of $r(z)$ in the proof above is certainly not the most practical one. Taking $r(z)$ as the least common multiple of the denominators $q_{ij}(z)$ is already better.

The foregoing result in terms of the transfer function has the following corollary for system representation in the time domain.

Corollary 2.3.5 *A causal linear time-invariant system Σ has a state space representation if and only if the input-output map is of polynomial form, that is,*

$$y(t) = A_1 y(t - 1) + \cdots + A_L y(t - L) + B_0 u(t) + B_1 u(t - 1) + \cdots + B_L u(t - L)$$

for $p \times p$ matrices A_i and $p \times q$ matrices B_j, $i = 1, \ldots, L$, $j = 0, \ldots, L$. Here we take $u(j) = 0$ for $j < 0$.

Proof This is an immediate consequence of Theorem 2.3.3. As we have seen in the proof of Theorem 2.3.3 a proper rational transfer function can be written as $\widehat{G}(z) = G(0) + \frac{H(z)}{r(z)}$. Denoting $L(z) = r(z)I$, we rewrite this as $\widehat{G}(z) = L(z)^{-1}(G(0)r(z) + H(z))$. We conclude that any proper rational transfer function can be written as the quotient of two matrix polynomials: $\widehat{G}(z) = \widehat{A}^{-1}(z)\widehat{B}(z)$ with $\widehat{A}(z) = I - A_1 z^{-1} - \cdots - A_L z^{-L}$ and $\widehat{B}(z) = B_0 + B_1 z^{-1} + \cdots + B_L z^{-L}$. Observe that it is quite well possible that some of the B_j's are zero. Rewriting the relation $\widehat{y}(z) = \widehat{G}(z)\widehat{u}(z)$ as $\widehat{A}(z)\widehat{y}(z) = \widehat{B}(z)\widehat{u}(z)$, and transforming this relation to the time domain gives the result. Note that there may be many ways in which we can write $\widehat{G}(z)$ as $\widehat{A}^{-1}(z)\widehat{B}(z)$, i.e., the matrix polynomials $A(z)$ and $B(z)$ are not uniquely determined.

Conversely, suppose that $\widehat{A}(z)\widehat{y}(z) = \widehat{B}(z)\widehat{u}(z)$. As $\widehat{A}$ is proper, $\widehat{A}^{-1}(z)$ is rational and proper. Hence $\widehat{G}(z) = \widehat{A}^{-1}(z)\widehat{B}(z)$ is rational and proper as well. Consequently it has a state space representation. $\qquad \square$

2.4   Equivalent and Minimal Realizations

An ordered quadruple (A, B, C, D) is called a *realization* of the system Σ if

$$
\begin{cases}
x(t+1) = Ax(t) + Bu(t), \quad t = 0, 1, 2, \ldots, \\
y(t) = Cx(t) + Du(t)
\end{cases}
\tag{2.19}
$$

is a state space representation of the input-output map of Σ. In other words, (A, B, C, D) is a realization of Σ if and only if Σ has transfer function $\widehat{G}(z) = D + C(z - A)^{-1}B$.

Realizations are not unique. For example, we may carry out a basis transformation in the state space, replacing the state variable $x(t)$ in (2.19) by $\widetilde{x}(t) = Sx(t)$, where S is an invertible matrix. In terms of the state $\widetilde{x}(t)$, we have

$$
\widetilde{x}(t+1) = Sx(t+1) = SAx(t) + SBu(t) = SAS^{-1}\widetilde{x}(t) + SBu(t),
$$

and $y(t) = Cx(t) + Du(t) = CS^{-1}\widetilde{x}(t) + Du(t)$. One sees that $G(k) = CA^{k-1}B = CS^{-1}(SAS^{-1})^{k-1}SB$, so the impulse response does not change, and hence also the transfer function does not change. Thus if (A, B, C, D) is a realization of Σ, then the same holds true for $(SAS^{-1}, SB, CS^{-1}, D)$.

There is another source of non-uniqueness. Let (A_0, B_0, C_0, D_0) be a realization, and let

$$
A = \begin{pmatrix} A_1 & A_3 & A_4 \\ 0 & A_0 & A_5 \\ 0 & 0 & A_2 \end{pmatrix}, \quad B = \begin{pmatrix} B_1 \\ B_0 \\ 0 \end{pmatrix}, \quad C = \begin{pmatrix} 0 & C_0 & C_2 \end{pmatrix},
\tag{2.20}
$$

where $A_1, A_2, A_3, A_4, A_5, B_1$ and C_2 are free to choose. Then

$$
A^k = \begin{pmatrix} * & * & * \\ 0 & A_0^k & * \\ 0 & 0 & * \end{pmatrix},
$$

where the $*$'s denote entries which we do not specify further. It then follows that $CA^kB = C_0A_0^kB_0$ (for $k \geq 0$). So, if (A_0, B_0, C_0, D_0) is a realization of Σ then the same holds true for (A, B, C, D_0) (compare Theorem 2.3.1).

The above two operations describe all possible realizations, as will be shown in Chap. 3. Of course, the realization (2.20) is less attractive than the realization (A_0, B_0, C_0, D_0), as it involves more parameters and more state variables.

A realization (A, B, C, D) of a system Σ is called *minimal* if among all realizations of Σ the state space dimension (that is, the size of A) is as small as possible.

2.5   The Restricted Shift Realization

We shall now construct a minimal realization from the matrices in the impulse response sequence. The construction is done using as a starting point an infinite dimensional vector space, and some linear transformations acting on it. By $\mathcal{L}^m$ we denote the linear space of all sequences $u = (u(1), u(2), u(3), \ldots)$ with elements in $\mathbb{R}^m$, and similarly, by $\mathcal{L}^p$ we denote the space of all sequences $y = (y(1), y(2), y(3), \ldots)$ with elements in $\mathbb{R}^p$. By $\mathcal{L}_0^m$ we denote the subspace of $\mathcal{L}^m$ consisting of all sequences with *finite support*, that is, only a finite number of the $u(j)$'s are nonzero. Equivalently, $u(t) = 0$ for sufficiently large t. Similarly, $\mathcal{L}_0^p$ is the subspace of $\mathcal{L}^p$ consisting of all sequences of finite support. Let V be the linear transformation on $\mathcal{L}^p$, defined by $V(y(1), y(2), y(3), \ldots) = (y(2), y(3), y(4), \ldots)$. This linear transformation is called the *backward shift*. Observe that V leaves the space $\mathcal{L}_0^p$ invariant. By $H : \mathcal{L}_0^m \to \mathcal{L}^p$ we denote the linear transformation with matrix representation

$$H = \begin{pmatrix} G(1) & G(2) & G(3) \ldots \\ G(2) & G(3) & G(4) \ldots \\ G(3) & G(4) & G(5) \ldots \\ \vdots & \vdots & \vdots \end{pmatrix}. \tag{2.21}$$

That is, for $u = (u(1), u(2), \ldots) \in \mathcal{L}_0^m$ the sequence $y = Hu$ has i-th entry $y(i) = \sum_{j=1}^{\infty} G(i + j - 1)u(j)$. Since u has finite support the right-hand side is a finite sum that always converges. (If sequences in $\mathcal{L}^m$ are considered as infinite columns, then the action of H is given by the usual matrix multiplication.) The infinite matrix (2.21) is called a block-Hankel matrix, as it has constant values on the counter (block) diagonals. The range of the linear transformation H will play an important role in the next theorem, we shall denote it by $\operatorname{Im} H$.

Theorem 2.5.1 *Let Σ be a causal linear time-invariant system with $p \times m$ impulse response matrices $G(0), G(1), G(2), \ldots$. Let $H : \mathcal{L}_0^m \to \mathcal{L}^p$ be given by (2.21), and define $X = \operatorname{Im} H$. Let V denote the backward shift on $\mathcal{L}^p$. Then the minimal state space dimension of realizations of Σ is equal to the dimension of X. Furthermore, if $k = \dim X < \infty$ then a minimal realization is obtained by taking*

$$A = V|_X : X \to X,$$

$$B = \begin{pmatrix} G(1) \\ G(2) \\ \vdots \end{pmatrix} : \mathbb{R}^m \to \operatorname{Im} H, \tag{2.22}$$

$$C = \begin{pmatrix} I & 0 & 0 & \dots \end{pmatrix} |_{\operatorname{Im} H} : \operatorname{Im} H \to \mathbb{R}^p,$$

$$D = G(0) : \mathbb{R}^m \to \mathbb{R}^p.$$

Proof Let $\widetilde{\Theta} = (\widetilde{A}, \widetilde{B}, \widetilde{C}, \widetilde{D})$ be a realization of Σ with state space $\mathbb{R}^n$. Define

$$\widetilde{\Lambda} = \begin{pmatrix} \widetilde{B} & \widetilde{A}\widetilde{B} & \widetilde{A}^2\widetilde{B} & \dots \end{pmatrix} : \mathcal{L}_0^m \to \mathbb{R}^n,$$

and

$$\widetilde{\Gamma} = \begin{pmatrix} \widetilde{C} \\ \widetilde{C}\widetilde{A} \\ \widetilde{C}\widetilde{A}^2 \\ \vdots \end{pmatrix} : \mathbb{R}^n \to \mathcal{L}^p,$$

where sequences in $\mathcal{L}^m$ and $\mathcal{L}^p$ are written as infinite columns. As sequences in $\mathcal{L}_0^m$ have finite support, the operator $\widetilde{\Lambda}$ is well-defined. Since $\widetilde{\Theta}$ is a realization of Σ, the j-th value of the impulse response of Σ is given by $G(j) = \widetilde{C}\widetilde{A}^{j-1}\widetilde{B}$ for $j \geq 1$. It follows that $H = \widetilde{\Gamma}\widetilde{\Lambda}$, and hence $k = \dim \operatorname{Im} H \leq \dim \operatorname{Im} \widetilde{\Lambda} \leq n$. So if a realization exists, then $\dim \operatorname{Im} H < \infty$ and the state dimension is at least $k = \dim \operatorname{Im} H$. To complete the proof, it suffices to show that $\Theta = (A, B, C, D)$ as given by (2.22) is a realization of Σ. Let $X = \operatorname{Im} H$ be finite dimensional. As

$$VH = \begin{pmatrix} G(2) & G(3) \dots \\ G(3) & G(4) \dots \\ G(4) & G(5) \dots \\ \vdots & \vdots \end{pmatrix} : \mathcal{L}_0^m \to \mathcal{L}^p,$$

it follows that $VX = V(\operatorname{Im} H) = \operatorname{Im} VH \subset \operatorname{Im} H = X$, and hence X is invariant under V. Therefore A in (2.22) is a well-defined linear transformation. Since $\operatorname{Im} B \subset \operatorname{Im} H$, we have

$$A^{j-1}B = V^{j-1}B = \begin{pmatrix} G(j) \\ G(j+1) \\ \vdots \end{pmatrix}, \qquad j \geq 1.$$

This shows that $G(j) = CA^{j-1}B$ for $j \geq 1$, that is, $\Theta = (A, B, C, D)$ is a realization of Σ. $\square$

The realization $\Theta = (A, B, C, D)$ described in Theorem 2.5.1 is called a *restricted shift realization*. An algorithm to construct a minimal realization from the Hankel matrix H will be given in Sect. 3.4.

State Space Models

3

This chapter discusses structure theory of state space systems. The central concepts are observability (the possibility to reconstruct the state from inputs and outputs) and controllability (the possibility to influence the state by manipulating the inputs). Minimal realizations are observable and controllable, and the converse is also true. We characterize all non-minimal realizations, and give an algorithm to compute the matrices in a minimal realization from the impulse response of the system.

3.1 Controllability

A realization $\Theta = (A, B, C, D)$ of a system Σ is called *controllable* if, starting from an arbitrary initial state x_0, any other state x_1 can be reached in finite time by choosing an appropriate input sequence. To make this more precise, let $x(t; x_0, u)$ be the solution at time t of the recursion $x(k + 1) = Ax(k) + Bu(k)$, $k = 0, 1, 2, \ldots$, with $x(0) = x_0$, and input sequence u, so that

$$x(t; x_0, u) = A^t x_0 + \sum_{j=0}^{t-1} A^{t-1-j} Bu(j), \qquad t \geq 1. \tag{3.1}$$

Thus Θ is controllable if and only if for every $x_0, x_1 \in \mathbb{R}^n$ there exist $t > 0$ and an input sequence u such that $x_1 = x(t; x_0, u)$.

A related notion is that of reachability. A realization is called *reachable* if starting from the origin $x_0 = 0$ every other state can be reached with an appropriate input sequence in a finite time interval, so that for every $x \in \mathbb{R}^n$ there exist $t > 0$ and an input sequence u such that $x = x(t; 0, u)$. Obviously, a controllable realization is also reachable.

© The Author(s), under exclusive license to Springer Nature Switzerland AG 2021
C. Heij et al., *Introduction to Mathematical Systems Theory*,
https://doi.org/10.1007/978-3-030-59654-5_3

For the analysis of controllability and reachability it is helpful to consider the sets $\mathcal{R}_t(\Theta)$ of states that can be reached at time t with an appropriate input sequence $u(0), u(1), \ldots, u(t-1)$ starting from the origin, that is

$$\mathcal{R}_t(\Theta) = \{x \in \mathbb{R}^n \mid \text{there exists } u \text{ such that } x = x(t; 0, u)\}.$$

It follows from (3.1) with $x_0 = 0$ that $\mathcal{R}_t(\Theta)$ is the image of the partitioned matrix $\begin{pmatrix} B & AB & \ldots & A^{t-1}B \end{pmatrix}$, i.e.,

$$\mathcal{R}_t(\Theta) = \text{Im} \begin{pmatrix} B & AB & \ldots & A^{t-1}B \end{pmatrix}. \tag{3.2}$$

In the sequel we shall use the notation X^T for the transpose of a matrix X.

Theorem 3.1.1 *Let $\Theta = (A, B, C, D)$ be a realization of the system Σ with state space dimension n. Then the following statements are equivalent:*

  (i) *Θ is controllable,*

 (ii) *Θ is reachable,*

(iii) *rank $\begin{pmatrix} B & AB & \ldots & A^{n-1}B \end{pmatrix} = n$,*

(iv) *the matrix $\sum_{j=0}^{n-1} A^j BB^T (A^T)^j$ is non-singular.*

Proof (i) $\Rightarrow$ (ii). This follows directly from the definitions.

(ii) $\Rightarrow$ (iii). Let us denote by $p(z) = z^n + p_{n-1}z^{n-1} + \cdots + p_0$ the characteristic polynomial of A. By the Cayley-Hamilton theorem we have $p(A) = 0$, and so the matrix A^n is a linear combination of the matrices $I, A, \ldots, A^{n-1}$. It follows from this that also the matrix A^{n+k} ($k \geq 0$) is a linear combination of the matrices $I, A, \ldots, A^{n-1}$. Hence $\mathcal{R}_k(\Theta) = \mathcal{R}_n(\Theta)$ for all $k \geq n$. When Θ is reachable there holds $\mathbb{R}^n = \cup_{k \geq 1}\mathcal{R}_k(\Theta) = \mathcal{R}_n(\Theta) = \text{Im} \begin{pmatrix} B & AB & \ldots & A^{n-1}B \end{pmatrix}$, and this shows (iii).

(iii) $\Rightarrow$ (i). By assumption $\mathcal{R}_n(\Theta) = \mathbb{R}^n$, so that $x - A^n x_0 \in \mathcal{R}_n(\Theta)$ for every $x, x_0 \in \mathbb{R}^n$. So there exists an input sequence $u = (u(0), u(1), \ldots, u(n-1))$ with $x - A^n x_0 = \sum_{j=0}^{n-1} A^{n-1-j} Bu(j)$. Therefore, $x = x(n; x_0, u)$, and Θ is controllable.

(iii) $\Leftrightarrow$ (iv). Let $\Lambda = \begin{pmatrix} B & AB & \ldots & A^{n-1}B \end{pmatrix}$, then we have the equality $\Lambda\Lambda^T = \sum_{j=0}^{n-1} A^j BB^T (A^T)^j$, and the result follows from the fact that rank $\Lambda = $ rank $\Lambda\Lambda^T$, and the fact that the square $n \times n$ matrix $\Lambda\Lambda^T$ is non-singular if and only if its rank is n.   □

Note that controllability of the realization $\Theta = (A, B, C, D)$ is independent of the matrices C and D. For that reason, instead of saying that the realization $\Theta = (A, B, C, D)$ is controllable, this is also expressed by saying that the pair (A, B) is controllable. The condition (iv) shows that controllability is a "robust" property, as controllability is preserved under small perturbations of the system parameters A and B.

Example 3.1.1 Consider the state space representation (2.13) of the model of a national economy discussed in Example 2.3.1. Here the state space dimension is 2, and

$$A = \begin{pmatrix} \beta(1+\delta) & -\beta\delta \\ 1 & 0 \end{pmatrix}, \qquad B = \begin{pmatrix} 1 \\ 0 \end{pmatrix}.$$

Thus

$$\begin{pmatrix} B & AB \end{pmatrix} = \begin{pmatrix} 1 & \beta(1+\delta) \\ 0 & 1 \end{pmatrix},$$

which obviously has rank 2. So this realization is controllable.

In Sect. 1.4 we also considered a three-dimensional realization of this system, with the state space matrices (corresponding to the single output $y(t)$) given by

$$A = \begin{pmatrix} 0 & 0 & \beta \\ -\delta & 0 & \beta\delta \\ \delta & 0 & \beta(1+\delta) \end{pmatrix}, B = \begin{pmatrix} 0 \\ 0 \\ 1 \end{pmatrix}, C = \begin{pmatrix} -\delta & 0 & \beta(1+\delta) \end{pmatrix}, D = 1.$$

It is easily seen that this system is controllable provided $\beta \neq 0$ and $\delta \neq 0$. If for example $\delta = 0$, so that investments are constant, then

$$\begin{pmatrix} B & AB & A^2B \end{pmatrix} = \begin{pmatrix} 0 & \beta & \beta^2 \\ 0 & 0 & 0 \\ 1 & \beta & \beta^2 \end{pmatrix},$$

and the set of reachable states is given by

$$\mathcal{R} = \{ \begin{pmatrix} x_1 & x_2 & x_3 \end{pmatrix}^T \in \mathbb{R}^3 \mid x_2 = 0 \}.$$

So, if $\delta = 0$ then this state space system is not reachable.

3.2   Observability

A state space system is called *observable* if the state vector can be reconstructed from the inputs and outputs. By $y(t; x_0, u)$ we denote the output at time t generated by the input sequence u and initial state $x(0) = x_0$ in the system

$$\begin{cases} x(t+1) = Ax(t) + Bu(t), & t \geq 0, \\ y(t) = Cx(t) + Du(t). \end{cases} \tag{3.3}$$

In other words, $y(t; x_0, u) = Cx(t; x_0, u) + Du(t)$, where $x(t; x_0, u)$ is given by (3.1). The realization $\Theta = (A, B, C, D)$ is called *observable* if for some input sequence the following implication holds:

$$y(t; x_0, u) = y(t; \tilde{x}_0, u), \ t \geq 0 \Rightarrow x_0 = \tilde{x}_0. \tag{3.4}$$

This means that the initial state at time $t = 0$ is uniquely determined by the inputs and outputs. The particular choice of the input u is irrelevant here. Indeed, $x(t; x_0, u) = x(t; x_0, 0) + \sum_{j=0}^{t-1} A^{t-1-j} Bu(j)$, and hence $y(t; x_0, u) = y(t; x_0, 0) + \sum_{j=0}^{t-1} CA^{t-1-j} Bu(j) + Du(t)$, so (3.4) holds if and only if

$$y(t; x_0, 0) = y(t; \tilde{x}_0, 0), \ t \geq 0 \Rightarrow x_0 = \tilde{x}_0. \tag{3.5}$$

A state x in the state space $\mathbb{R}^n$ is said to be *unobservable* over the time interval $t = 0, \ldots, k - 1$ if $y(t; x, 0) = CA^t x = 0, \ t = 0, \ldots, k - 1$. The set of all states that are unobservable over $t = 0, \ldots, k - 1$ is denoted by

$$\mathcal{N}_k(\Theta) = \{x \in \mathbb{R}^n \mid y(t; x, 0) = 0 \text{ for } t = 0, \ldots, k - 1\}.$$

It follows that

$$\mathcal{N}_k(\Theta) = \text{Ker} \begin{pmatrix} C \\ CA \\ \vdots \\ CA^{k-1} \end{pmatrix}. \tag{3.6}$$

Theorem 3.2.1 *Let* $\Theta = (A, B, C, D)$ *be a realization of the system* Σ *with state space dimension n. Then the following statements are equivalent:*

(i) Θ *is observable,*

(ii) rank $\begin{pmatrix} C \\ CA \\ \vdots \\ CA^{n-1} \end{pmatrix} = n,$

(iii) *the matrix* $\sum_{j=0}^{n-1} (A^T)^j C^T C A^j$ *is non-singular.*

Proof (i) $\Leftrightarrow$ (ii). By the Cayley-Hamilton theorem, the matrix A^t $(t \geq n)$ is a linear combination of the matrices $I, A, \ldots, A^{n-1}$. This means that $\mathcal{N}_t(\Theta) = \mathcal{N}_n(\Theta)$ for all $t \geq n$. According to (3.5), the realization is observable if and only if

$$\{0\} = \cap_{k \geq 1} \mathcal{N}_k(\Theta),$$

which in turn is equivalent to

$$\{0\} = \mathcal{N}_n(\Theta).$$

The result now follows from (3.6).

(ii) $\Leftrightarrow$ (iii). This follows from the fact that rank $M =$ rank $M^T M$, and the matrix $M^T M$ is non-singular if and only if its rank is n. $\square$

The equivalence of (i) and (iii) in Theorem 3.2.1 shows that observability is preserved under small perturbations of the system parameters. Since the condition for observability does not involve the matrices B and D, one says that (A, C) is observable if the realization $\Theta = (A, B, C, D)$ is observable.

From Theorem 3.1.1 (iv) and Theorem 3.2.1 (iii) we see that there is a kind of duality between observability and controllability, as (A, C) is observable if and only if (A^T, C^T) is controllable.

Example 3.2.1 Consider the realization (2.13) given in Example 2.3.1. The state space dimension is $n = 2$, and

$$A = \begin{pmatrix} \beta(1+\delta) & -\beta\delta \\ 1 & 0 \end{pmatrix}, \qquad C = \begin{pmatrix} \beta(1+\delta) & -\beta\delta \end{pmatrix},$$

so that

$$\begin{pmatrix} C \\ CA \end{pmatrix} = \begin{pmatrix} \beta(1+\delta) & -\beta\delta \\ \beta^2(1+\delta)^2 - \beta\delta & -\beta(1+\delta)\beta\delta \end{pmatrix}.$$

This matrix has rank 2 if and only if $\beta\delta \neq 0$, so the realization is observable if and only if $\beta \neq 0$ and $\delta \neq 0$. In our model of a national economy we may assume that $\beta > 0$ and $\delta > 0$, and hence for that case the system observable.

Next we consider the three dimensional realization of this system discussed in Sect. 1.4 with

$$A = \begin{pmatrix} 0 & 0 & \beta \\ -\delta & 0 & \beta\delta \\ \delta & 0 & \beta(1+\delta) \end{pmatrix}, \qquad C = \begin{pmatrix} -\delta & 0 & \beta(1+\delta) \end{pmatrix}.$$

This gives

$$\begin{pmatrix} C \\ CA \\ CA^2 \end{pmatrix} = \begin{pmatrix} * & 0 & * \\ * & 0 & * \\ * & 0 & * \end{pmatrix},$$

where $*$'s denote entries that depend on β and δ. Clearly, this matrix has rank 2 at most, so that this realization is not observable. In particular, it is not possible to reconstruct the second state variable.

This result may be understood by considering the model in Sect. 1.4 in more detail. If we apply the input $g(t) = 0$, $t \geq 0$, then the output $y(t)$, $t \geq 0$ can be used to reconstruct the initial values $y(-1)$ and $y(-2)$ in (1.33). It follows from (1.30) and (1.31) that also $c(t)$, $t \geq -1$ and $i(t)$, $t \geq 0$, can be derived from this information. However, the state $x(t) = \begin{pmatrix} c(t-1) & i(t-1) & y(t-1) \end{pmatrix}^T$ at time $t = 0$ also contains $i(-1)$, and this can not be calculated from the information in $y(t)$, $t \geq 0$. Therefore, this state component is not observable.

Another characterization of observability can be given in terms of the eigenvalues of the matrix A. Let $\Theta = (A, B, C, D)$, and let $\mathcal{N}(\Theta)$ be the subspace of unobservable states,

$$\mathcal{N}(\Theta) := \cap_{k\geq 1}\mathcal{N}_k(\Theta) = \cap_{k\geq 1}\text{Ker}\, CA^{k-1}.$$

Note that $\mathcal{N}(\Theta) = \mathcal{N}_n(\Theta)$ by the Cayley-Hamilton theorem. This subspace is invariant under A, that is, if $x \in \mathcal{N}(\Theta)$, then also $Ax \in \mathcal{N}(\Theta)$. We denote by $A_\mathcal{N}$ the restriction of A to $\mathcal{N}(\Theta)$ viewed as a map from $\mathcal{N}(\Theta)$ to itself. If $\mathcal{N}(\Theta) \neq \{0\}$, decompose $\mathbb{R}^n$ as $\mathbb{R}^n = \mathcal{N}(\Theta)\oplus\mathcal{N}(\Theta)^\perp$. (Here and in the sequel the symbol $\oplus$ denotes the orthogonal direct sum of two subspaces.) Choosing bases in $\mathcal{N}(\Theta)$ and $\mathcal{N}(\Theta)^{\perp}$, and combining these bases to a basis for $\mathbb{R}^n$, we can write a matrix for A with respect to this basis. The invariance of $\mathcal{N}(\Theta)$ implies that this matrix for A has the form

$$A = \begin{pmatrix} A_{11} & A_{12} \\ 0 & A_{22} \end{pmatrix}. \tag{3.7}$$

Here, A_{11} is a matrix representation of $A_{\mathcal{N}}$. So, if $\mathcal{N}(\Theta) \neq \{0\}$, then $A_{\mathcal{N}}$ has eigenvalues. Every eigenvalue of $A_{\mathcal{N}}$ is also an eigenvalue of A. An eigenvalue of A is called a (A, C)-*observable* if it is not an eigenvalue of $A_{\mathcal{N}}$. The (A, C)-*unobservable eigenvalues* are defined as the eigenvalues of A that are also eigenvalues of $A_{\mathcal{N}}$. Hence, in the representation (3.7) the (A, C)-unobservable eigenvalues are those of A_{11}, while the observable eigenvalues are those of A_{22} that are not also eigenvalues of A_{11}.

Theorem 3.2.2 *Let* $\Theta = (A, B, C, D)$ *be a realization with state space dimension* n. *Then the following statements are equivalent:*

(i) Θ *is observable,*

(ii) $\operatorname{rank} \begin{pmatrix} A - \lambda I \\ C \end{pmatrix} = n$ *for each* $\lambda \in \mathbb{C}$,

(iii) $\operatorname{rank} \begin{pmatrix} A - \lambda I \\ C \end{pmatrix} = n$ *for each eigenvalue* λ *of* A,

(iv) *all eigenvalues of* A *are* (A, C) *observable.*

Proof Let $M(\lambda) = \begin{pmatrix} A - \lambda I \\ C \end{pmatrix} : \mathbb{C}^n \to \mathbb{C}^{n+p}$. It is important here that we consider this as a map between complex vector spaces, as the eigenvalues of the (real) matrix A may be complex, and its eigenvectors may be complex vectors.

(ii)$\Leftrightarrow$(iii) If λ is not an eigenvalue of A then $M(\lambda)$ has rank n, so this equivalence is trivial.

(iii)$\Leftrightarrow$(iv) Suppose rank $M(\lambda) < n$. Then there exists a (possibly complex) vector $x \neq 0$ with $M(\lambda)x = 0$, so that $Ax = \lambda x$ and $Cx = 0$. This implies that $CA^{t-1}x = \lambda^{t-1}Cx = 0$ for $t \geq 0$, and the same holds true for the real and imaginary parts of the vector x. This shows that λ is also an eigenvalue of $A_{\mathcal{N}}$, so that it is an (A, C)-unobservable eigenvalue. Conversely, if λ is an (A, C)-unobservable eigenvalue then there exists $x \neq 0$ with $Ax = \lambda x$ and $x \in \mathcal{N}(\Theta) + i\mathcal{N}(\Theta)$. (Observe that $\mathcal{N}(\Theta)$ is a real vector space.) This implies that $Cx = 0$, so that $M(\lambda)x = 0$, and hence rank $M(\lambda) < n$.

(i)⇔(iv) If Θ is observable then $\mathcal{N}(\Theta) = \{0\}$ by definition, and hence there can be no (A, C)-unobservable eigenvalues. On the other hand, if all eigenvalues of A are (A, C)-observable, this means that $\mathcal{N}(\Theta) = \{0\}$ so that Θ is observable.   □

The condition (iii) is called the *Hautus test* for observability.

By using transposition and the remark preceding Example 3.2.1, one obtains the notions of controllable and uncontrollable eigenvalues. An eigenvalue λ of A is called an (A, B)-*controllable eigenvalue* if rank $\begin{pmatrix} A - \lambda I & B \end{pmatrix} = n$. Otherwise it is called an *uncontrollable eigenvalue*. For example, if $A = \begin{pmatrix} \frac{1}{2} & 0 \\ 0 & \frac{1}{2} \end{pmatrix}$ and $B = \begin{pmatrix} 1 \\ 0 \end{pmatrix}$, then $\frac{1}{2}$ is not a controllable eigenvalue. The following Hautus test for controllability is an immediate corollary of Theorem 3.2.2.

Theorem 3.2.3 *Let* $\Theta = (A, B, C, D)$ *be a realization with state space dimension n. Then the following statements are equivalent:*

(i) Θ *is controllable,*
(ii) rank $\begin{pmatrix} A - \lambda I & B \end{pmatrix} = n$ *for each* $\lambda \in \mathbb{C}$,
(iii) rank $\begin{pmatrix} A - \lambda I & B \end{pmatrix} = n$ *for each eigenvalue* λ *of A,*
(iv) *all eigenvalues of A are* (A, B) *controllable.*

3.3   Structure Theory of Realizations

In this section we describe the structure of state space representations of a given system. We pay particular attention to *minimal realizations*, that is, realizations with the lowest state space dimension. As a first step, we show that realizations that are uncontrollable or unobservable can be reduced to realizations of smaller dimension. For this purpose we use the following terminology.

Two realizations $\Theta = (A, B, C, D)$ and $\Theta_0 = (A_0, B_0, C_0, D_0)$ are called *similar* if (i) $D = D_0$, (ii) Θ and Θ_0 have the same state space $\mathbb{R}^n$ and (iii) there exists an invertible linear transformation $S : \mathbb{R}^n \to \mathbb{R}^n$ such that $A = SA_0S^{-1}$, $B = SB_0$, $C = C_0S^{-1}$.

The realization Θ is called a *dilation* of Θ_0 or equivalently, Θ_0 is a *reduction* of Θ, if (i) $D = D_0$, and (ii) for a suitable choice of $A_1, A_2, A_3, A_4, A_5, B_1$ and C_2 the three identities in (2.20) hold true, that is,

$$A = \begin{pmatrix} A_1 & A_3 & A_4 \\ 0 & A_0 & A_5 \\ 0 & 0 & A_2 \end{pmatrix}, \quad B = \begin{pmatrix} B_1 \\ B_0 \\ 0 \end{pmatrix}, \quad C = \begin{pmatrix} 0 & C_0 & C_2 \end{pmatrix}.$$

It is an easy exercise to show that similar realizations produce the same input-output behavior, provided both are started with initial state zero. Likewise, if Θ is a dilation of Θ_0, then these two realizations produce the same input-output behavior when started with initial state zero.

Let $\Theta = (A, B, C, D)$ be a realization of the system Σ, and let the state space dimension of this realization be n. The reachable subspace associated with Θ is denoted by $\mathcal{R}(\Theta)$ and the unobservable subspace by $\mathcal{N}(\Theta)$. Thus

$$\mathcal{R}(\Theta) = \mathrm{Im}\left(B \ AB \ \ldots \ A^{n-1}B\right),$$

$$\mathcal{N}(\Theta) = \bigcap_{k=0}^{n-1} \mathrm{Ker}\, CA^k.$$

Observe that $\mathcal{R}(\Theta) = \mathrm{Im}\left(B \ AB \ A^2B \ \ldots\right)$ by the Cayley-Hamilton theorem. The state space is decomposed as a direct sum as

$$\mathbb{R}^n = X_1 \dot{+} X_2 \dot{+} X_3 \dot{+} X_4, \tag{3.8}$$

where $X_1 = \mathcal{N}(\Theta) \sqcap \mathcal{R}(\Theta)$, $X_1 \dot{+} X_2 = \mathcal{N}(\Theta)$, $X_1 \dot{+} X_3 = \mathcal{R}(\Theta)$, and $\{\mathcal{N}(\Theta) + \mathcal{R}(\Theta)\} \dot{+} X_4 = \mathbb{R}^n$. (Here and in the sequel $\dot{+}$ denotes the direct sum of subspaces.) Let X_i have dimension n_i, $i = 1, 2, 3, 4$, so that $n_1 + n_2 + n_3 + n_4 = n$, and let $b_1, \ldots, b_n$ be a basis for $\mathbb{R}^n$ ordered in such a way that the first n_1 vectors are a basis for X_1, the next n_2 vectors are a basis for X_2, the next n_3 vectors from a basis of X_3, and finally, the last n_4 vectors are a basis for X_4. Let $S = \left(b_1 \ \ldots \ b_n\right)$, so that S is invertible, and let

$$\widetilde{\Theta} := (S^{-1}AS, S^{-1}B, CS, D). \tag{3.9}$$

As Θ and $\widetilde{\Theta}$ are similar, both are realizations of Σ. The matrices $S^{-1}AS$, $S^{-1}B$ and CS have a special structure, namely

$$S^{-1}AS = \begin{pmatrix} A_{11} & A_{12} & A_{13} & A_{14} \\ 0 & A_{22} & 0 & A_{24} \\ 0 & 0 & A_{33} & A_{34} \\ 0 & 0 & 0 & A_{44} \end{pmatrix},$$

$$S^{-1}B = \begin{pmatrix} B_1 \\ 0 \\ B_3 \\ 0 \end{pmatrix}, \qquad CS = \begin{pmatrix} 0 & 0 & C_3 & C_4 \end{pmatrix}$$

The partitioning is in accordance with the above decomposition of the state space, and we have used that $A\mathcal{N}(\Theta) \subset \mathcal{N}(\Theta)$, $A\mathcal{R}(\Theta) \subset \mathcal{R}(\Theta)$, Im $B \subset \mathcal{R}(\Theta)$ and $\mathcal{N}(\Theta) \subset$ Ker C. It follows that $\widetilde{\Theta}$ is a dilation of $\Theta_0 = (A_{33}, B_3, C_3, D)$.

Proposition 3.3.1 *The quadruple* $\Theta_0 = (A_{33}, B_3, C_3, D)$ *is a controllable and observable realization of* Σ.

Proof As $\widetilde{\Theta}$ is a dilation of Θ_0 it is easily checked that they have the same impulse response. Therefore they represent the same system Σ.

The realization Θ_0 is controllable. If this were not the case, then A_{33} has an (A_{33}, B_3)-uncontrollable eigenvalue λ. Thus there exists a vector $x_3 \neq 0$ in $\mathbb{C}^{n_3}$ such that

$$x_3^* A_{33} = \lambda x_3^*, \qquad x_3^* B_3 = 0.$$

Let $x = \left(0 \; 0 \; x_3^T \; 0\right)^T$, then $x^*(S^{-1}AS)^k S^{-1} B = 0$, $k = 0, 1, 2, \ldots$, so that x is orthogonal to $S^{-1}\mathcal{R}(\Theta)$. As on the other hand $Sx \in X_3 \subset \mathcal{R}(\Theta)$ it follows that $x = 0$, and hence $x_3 = 0$. This shows that Θ_0 is controllable.

The realization Θ_0 is also observable. Let $x_3 \in \mathcal{N}(\Theta_0)$, so $C_3 A_{33}^k x_3 = 0$ for each $k \geq 0$. Hence $CS(S^{-1}AS)^k x = 0$, $k \geq 0$, where $x = \left(0 \; 0 \; x_3^T \; 0\right)^T$. This means that $Sx \in \mathcal{N}(\Theta) = X_1 \dot{+} X_2$, but also $Sx \in X_3$ and therefore $x = 0$ and thus also $x_3 = 0$. $\square$

Proposition 3.3.2 *Two controllable and observable realizations of the same system* Σ *are similar, and the corresponding state space similarity transformation is unique.*

Proof Let $\Theta_1 = (A_1, B_1, C_1, D)$ and $\Theta_2 = (A_2, B_2, C_2, D)$ be controllable and observable realizations of Σ with state space dimensions n_1 and n_2, respectively. Let $G(\cdot)$ be the impulse response matrix of Σ, then

$$G(j) = C_1 A_1^{j-1} B_1 = C_2 A_2^{j-1} B_2, \qquad j \geq 1. \tag{3.10}$$

Let n be the largest of the two numbers n_1 and n_2, and let H_n be the block Hankel matrix defined by

$$H_n = \begin{pmatrix} G(1) & G(2) & \ldots & G(n) \\ G(2) & G(3) & \ldots & G(n+1) \\ \vdots & \vdots & & \vdots \\ G(n) & G(n+1) & \ldots & G(2n-1) \end{pmatrix}.$$

It follows from (3.10) that $H_n = \Gamma_1(n)\Lambda_1(n) = \Gamma_2(n)\Lambda_2(n)$, where

$$
\Gamma_i(n) = \begin{pmatrix} C_i \\ C_i A_i \\ \vdots \\ C_i A_i^{n-1} \end{pmatrix}, \quad \Lambda_i(n) = \begin{pmatrix} B_i & A_i B_i & \ldots & A_i^{n-1} B_i \end{pmatrix}, \quad i = 1, 2. \tag{3.11}
$$

As Θ_1 and Θ_2 are controllable and observable and $n \geq n_i, i = 1, 2$, it follows that $\operatorname{Im} \Lambda_i(n) = \mathbb{R}^{n_i}$ and $\operatorname{Ker} \Gamma_i(n) = \{0\}$, so that rank $H_n = n_1 = n_2 = n$.

To prove the similarity of Θ_1 and Θ_2, we define $S : \mathbb{R}^n \to \mathbb{R}^n$ as follows. For every $x \in \mathbb{R}^n$ and every $k \geq n$ there exist $u_j \in \mathbb{R}^m$, $j = 0, \ldots, k-1$, so that

$$
x = \sum_{j=0}^{k-1} A_1^j B_1 u_j. \tag{3.12}
$$

We define $Sx = \sum_{j=0}^{k-1} A_2^j B_2 u_j$. This definition does not depend on the particular choice of the vectors $u_0, \ldots, u_{k-1}$. Indeed, let $k' \geq n$ and $x = \sum_{j=0}^{k'-1} A_1^j B_1 u'_j$. By adding zero vectors if necessary we may assume without loss of generality that $k = k'$. Let u and u' be the vectors in $(\mathbb{R}^m)^k$ with components u_j and u'_j, respectively. Then $H_k(u - u') = \Gamma_1(k)\Lambda_1(k)(u - u') = \Gamma_1(k)(x - x) = 0$. As $H_k = \Gamma_2(k)\Lambda_2(k)$ it follows that $\Gamma_2(k)\Lambda_2(k)(u - u') = 0$, and as Θ_2 is observable this implies that $\Lambda_2(k)(u - u') = 0$, that is $\sum_{j=0}^{k-1} A_2^j B_2 u_j = \sum_{j=0}^{k-1} A_2^j B_2 u'_j$. This shows that S is well-defined.

It is straightforward to check that S is a linear operator that is surjective, as $\operatorname{Im} S = \operatorname{Im} \Lambda_2(n) = \mathbb{R}^n$ because Θ_2 is controllable. Therefore S is invertible. From the definition of S it follows that $SB_1 u = B_2 u$ for every $u \in \mathbb{R}^m$, so that $SB_1 = B_2$. Further, for x as in (3.12) there holds

$$
S(A_1 x) = S\left(\sum_{j=0}^{k-1} A_1^{j+1} B_1 u_j \right) = \sum_{j=0}^{k-1} A_2^{j+1} B_2 u_j = A_2\left(\sum_{j=0}^{k-1} A_2^j B_2 u_j \right) = A_2 Sx.
$$

As Θ_1 is controllable this implies that $SA_1 = A_2 S$. Finally, for x as in (3.12) it follows from (3.10) that

$$
C_2 Sx = \sum_{j=0}^{k-1} C_2 A_2^j B_2 u_j = \sum_{j=0}^{k-1} C_1 A_1^j B_1 u_j = C_1 x,
$$

so that $C_2 S = C_1$. We have proved that $A_2 = S A_1 S^{-1}$, $B_2 = S B_1$, $C_2 = C_1 S^{-1}$, and hence Θ_1 and Θ_2 are similar.

To prove the uniqueness of S, let $\widetilde{S} : \mathbb{R}^n \to \mathbb{R}^n$ be invertible with $A_2 = \widetilde{S} A_1 \widetilde{S}^{-1}$, $B_2 = \widetilde{S} B_1$, $C_2 = C_1 \widetilde{S}^{-1}$, Then $C_2 A_2^j S = C_1 A_1^j = C_2 A_2^j \widetilde{S}$, $j \geq 0$, and as (A_2, C_2) is observable this implies that $S = \widetilde{S}$. $\qquad\square$

We now come to the two central results of realization theory.

Theorem 3.3.3 *A realization is minimal if and only if it is controllable and observable.*

Proof First we prove that a minimal realization is observable and controllable. Let Θ be a realization of the system Σ with state space $\mathbb{R}^n$, and suppose that Θ is not observable or not controllable. In terms of the decomposition (3.8) of the state space, this means that $\mathcal{N}(\Theta) = X_1 \dotplus X_2 \neq \{0\}$, or $\mathcal{R}(\Theta) \neq \mathbb{R}^n$ so that $X_4 \neq \{0\}$. Therefore $\dim X_3 < n$, and the realization of Proposition 3.3.1 has dimension smaller than n. This shows that Θ is not minimal.

To prove the converse, let Θ be a controllable and observable realization of Σ and let Θ_0 be a minimal realization of Σ, so that Θ_0 is also controllable and observable. Proposition 3.3.2 shows that Θ and Θ_0 are similar, so that the state space dimensions of Θ and Θ_0 are equal. Therefore Θ is also minimal. $\qquad\square$

Theorem 3.3.4 (i) *Two minimal realizations of Σ are similar, and the corresponding similarity transformation is unique.*
(ii) *Every realization of Σ is similar to a dilation of a minimal realization.*

Proof (i) This follows directly from Theorem 3.3.3 and Proposition 3.3.2.
(ii) Let Θ be a realization of Σ with state space $\mathbb{R}^n = X_1 \dotplus X_2 \dotplus X_3 \dotplus X_4$, decomposed as in (3.8). Then Θ is similar to the realization $\widetilde{\Theta}$ in (3.9), and $\widetilde{\Theta}$ is a dilation of the controllable and observable realization Θ_0 in Proposition 3.3.1. Theorem 3.3.3 implies that Θ_0 is a minimal realization. $\qquad\square$

3.4   An Algorithm for Minimal Realizations

In this section we present a matrix algorithm to construct minimal realizations for systems Σ with rational transfer function. The impulse response of Σ is denoted by $G(\cdot)$, and for every $k \geq 1$ we define the $pk \times mk$ block Hankel matrix

$$
H_k = \begin{pmatrix}
G(1) & G(2) & \ldots & G(k) \\
G(2) & G(3) & \ldots & G(k+1) \\
\vdots & \vdots & & \vdots \\
G(k) & G(k+1) & \ldots & G(2k-1)
\end{pmatrix}
$$

The result in Theorem 2.5.1 shows that the minimal state dimension of realizations of Σ is given by $n = \max_{k \geq 1} \operatorname{rank} H_k$. Let Θ_0 be a minimal realization, for instance the one constructed in Theorem 2.5.1. Then $H_n = \Gamma(n)\Lambda(n)$ where $\Gamma(n)$ and $\Lambda(n)$ are defined in (3.11), and $\operatorname{Im} \Lambda(n) = \mathbb{R}^n$, $\operatorname{Ker} \Gamma(n) = \{0\}$. This means that rank $H_n = n$, a result we shall use shortly in the algorithm. In what follows we assume that $n > 0$, as the case $n = 0$ is trivial. The following steps provide an algorithm to construct a minimal realization from a given impulse response.

Step 1.   Determine $n = \max_{k \geq 1} \operatorname{rank} H_k$, and recall that rank $H_n = n$.

Step 2.   Construct a minimal rank decomposition of H_n, that is, a factorization $H_n = \Gamma\Lambda$ where Γ is a $pn \times n$ matrix and Λ is a $n \times mn$ matrix. Let

$$
\Lambda = \begin{pmatrix} \Lambda_1 \ldots \Lambda_n \end{pmatrix}, \quad \Gamma = \begin{pmatrix} \Gamma_1 \\ \vdots \\ \Gamma_n \end{pmatrix},
$$

where Λ_j are $n \times m$ matrices and Γ_j are $p \times n$ matrices, for $j = 1, \ldots, n$. Define $B = \Lambda_1$ and $C = \Gamma_1$, and let $D = G(0)$.

Step 3.   Determine a right inverse Λ^+ of Λ and a left inverse of Γ^+ of Γ, and define

$$
A = \Gamma^+ \begin{pmatrix}
G(2) & G(3) & \ldots G(n+1) \\
G(3) & G(4) & \ldots G(n+2) \\
\vdots & \vdots & \vdots \\
G(n+1) & G(n+2) & \ldots & G(2n)
\end{pmatrix} \Lambda^+.
$$

Theorem 3.4.1 *The realization $\Theta = (A, B, C, D)$ constructed in the three steps above is a minimal realization of Σ.*

Proof   Let $\widetilde{\Theta} = (\widetilde{A}, \widetilde{B}, \widetilde{C}, \widetilde{D})$ be a minimal realization of Σ. According to Theorem 2.5.1 $\widetilde{\Theta}$ has state space dimension equal to n. Clearly, $\widetilde{D} = G(0) = D$, and as $G(j) = \widetilde{C}\widetilde{A}^{j-1}\widetilde{B}$

there holds $H_n = \widetilde{\Gamma}\widetilde{\Lambda}$, with $\widetilde{\Gamma}$ and $\widetilde{\Lambda}$ defined as in (3.11) in terms of $\widetilde{A}, \widetilde{B}, \widetilde{C}$. Then $H_n = \widetilde{\Gamma}\widetilde{\Lambda} = \Gamma\Lambda$ are two minimal rank decompositions, so there exists an invertible $n \times n$ matrix S such that $\widetilde{\Gamma} = \Gamma S$ and $\widetilde{\Lambda} = S^{-1}\Lambda$. Comparing Λ and Γ with $\widetilde{\Lambda}$ and $\widetilde{\Gamma}$ it follows that $CS = \widetilde{C}$ and $S^{-1}B = \widetilde{B}$. Since $H_n = \Gamma\Lambda$ is a minimal rank decomposition of H_n, it follows that Γ has a left inverse Γ^+ and Λ has a right inverse Λ^+. At this point it is instructive to note that $\widetilde{\Gamma}\widetilde{A}\widetilde{\Lambda}$ is equal to the matrix

$$\begin{pmatrix} G(2) & G(3) & \dots & G(n+1) \\ G(3) & G(4) & \dots & G(n+2) \\ \vdots & \vdots & & \vdots \\ G(n+1) & G(n+2) & \dots & G(2n) \end{pmatrix}.$$

Thus, the matrix A in step 3 is given by $A = \Gamma^+\widetilde{\Gamma}\widetilde{A}\widetilde{\Lambda}\Lambda^+ = \Gamma^+\Gamma S\widetilde{A}S^{-1}\Lambda\Lambda^+ = S\widetilde{A}S^{-1}$. It follows that $\Theta = (S\widetilde{A}S^{-1}, S\widetilde{B}, \widetilde{C}S^{-1}, \widetilde{D})$, and hence Θ is also a minimal realization.   □

To apply this algorithm we have to construct the factorization $H_n = \Gamma\Lambda$ and the matrices Γ^+ and Λ^+. This can be done in many ways. An explicit and convenient construction is as follows. The singular value decomposition of H_n gives

$$H_n = U\begin{pmatrix} \widehat{D} & 0 \\ 0 & 0 \end{pmatrix}V^T,$$

where U and V are orthogonal matrices and $\widehat{D}$ is an invertible $n \times n$ diagonal matrix. (See [20].) We rewrite this as

$$H_n = U\begin{pmatrix} \widehat{D} \\ 0 \end{pmatrix}\begin{pmatrix} I_n & 0 \end{pmatrix}V^T.$$

Now define $\Gamma = U\begin{pmatrix} \widehat{D} \\ 0 \end{pmatrix}$ and $\Lambda = \begin{pmatrix} I_n & 0 \end{pmatrix}V^T$. Clearly $H_n = \Gamma\Lambda$, and it is easy to see that this is a minimal rank decomposition of H_n. The matrices B and C are simply read off from Γ and Λ. In step 3 we can take $\Gamma^+ = \begin{pmatrix} \widehat{D}^{-1} & 0 \end{pmatrix}U^T$ and $\Lambda^+ = V\begin{pmatrix} I_n \\ 0 \end{pmatrix}$.

Example 3.4.1 As an illustration, we consider again the example of a national economy described by Eq. (2.12) and the minimal state space model (2.13). We assume that $\beta = 0.5$ and $\delta = 1$ in this model. Using (2.13) and Theorem 2.3.1, it follows that the impulse response of this system is given by $G(0) = 1$, $G(1) = 1$, $G(2) = 0.5$, $G(3) = 0$, $G(4) = -0.25$, and $G(4k+j) = (-0.25)^k G(j)$ for all $k \geq 1$ and $j = 1, 2, 3, 4$.

Step 1 of the algorithm gives $n = 2$ and

$$H_2 = \begin{pmatrix} 1 & 0.5 \\ 0.5 & 0 \end{pmatrix}.$$

For the rank decomposition we could of course use the singular value decomposition. However, in this case it can be taken very simple, as the matrix H_2 is invertible. So, we can use $\Gamma = H_2$ and $\Lambda = I_2$. Using this minimal rank decomposition we see that B is the first column of I_2 and C is the first row of H_2, that is, $B = \begin{pmatrix} 1 \\ 0 \end{pmatrix}$ and $C = \begin{pmatrix} 1 & 0.5 \end{pmatrix}$. Finally,

$$A = H_2^{-1} \begin{pmatrix} G(2) & G(3) \\ G(3) & G(4) \end{pmatrix} = \begin{pmatrix} 0 & -0.5 \\ 1 & 1 \end{pmatrix}.$$

In Examples 3.1.1 and 3.2.1 we considered the observable and controllable realization given by

$$A_0 = \begin{pmatrix} 1 & -0.5 \\ 1 & 0 \end{pmatrix}, \; B_0 = \begin{pmatrix} 1 \\ 0 \end{pmatrix}, \; C = \begin{pmatrix} 1 & -0.5 \end{pmatrix}.$$

It is easily verified that $(A, B, C) = (SA_0S^{-1}, SB_0, C_0S^{-1})$ with $S = \begin{pmatrix} 1 & -1 \\ 0 & 1 \end{pmatrix}$. It follows that (A, B, C) as constructed by the algorithm is indeed a minimal realization.

3.5   The Subspace Identification Algorithm

Assume that we have given a linear system

$$x_{k+1} = Ax_k + Bu_k,$$

$$y_k = Cx_k + Du_k,$$

with $u_k \in \mathbb{R}^m$, $x_k \in \mathbb{R}^n$ and $y_k \in \mathbb{R}^\ell$ and $x_0 = 0$. Suppose that the sequences $(u_k)_k$ and $(y_k)_k$ are given and that we want to reconstruct the system matrices (A, B, C, D) from these data. One should not expect to find a unique solution since for any invertible Q the quadruple $(Q^{-1}AQ, Q^{-1}B, QC, D)$ will generate the same outputs from the inputs as does the system (A, B, C, D).

The results in this section are based upon the subspace identification algorithm as it is described for a more general class of system in [55].

For the results that we are going to describe we need two integers, i and j, with i larger than n and j much larger than i. From the data we construct a number of matrices. We assume $k \leq p$ to be integers and put

$$
U_{k|p} = \begin{pmatrix} u_k & u_{k+1} & \cdots & u_{k+j-1} \\ u_{k+1} & u_{k+2} & & u_{k+j} \\ \vdots & & & \vdots \\ u_p & u_{p+1} & \cdots & u_{p+j-1} \end{pmatrix}, \quad Y_{k|p} = \begin{pmatrix} y_k & y_{k+1} & \cdots & y_{k+j-1} \\ y_{k+1} & y_{k+2} & & y_{k+j} \\ \vdots & & & \vdots \\ y_p & y_{p+1} & \cdots & y_{p+j-1} \end{pmatrix},
$$

$$
H_{k|k+i-1} = \begin{pmatrix} U_{k|k+i-1} \\ Y_{k|k+i-1} \end{pmatrix}.
$$

Note that $H_{k|k+i-1}$ is a $(\ell + m)i \times j$ matrix.

Suppose for a moment that we have a realization (A, B, C, D). Then we also can construct the sequence of states $(x_k)_k$. We define the $n \times j$ matrix

$$
X_k = \begin{pmatrix} x_k & x_{k+1} & \cdots & x_{k+j-1} \end{pmatrix}.
$$

By Sect. 2.3, see formula (2.14) and following, we have the relation

$$
Y_{k|k+i-1} = \Gamma X_k + H_T U_{k|k+i-1}, \tag{3.1}
$$

with

$$
\Gamma = \begin{pmatrix} C \\ CA \\ \vdots \\ CA^{i-1} \end{pmatrix}, \quad H_T = \begin{pmatrix} D & 0 & 0 & \cdots & 0 \\ CB & D & 0 & \cdots & 0 \\ CAB & CB & D & \ddots & \vdots \\ \vdots & \vdots & \ddots & \ddots & 0 \\ CA^{i-2}B & CA^{i-3}B & \cdots & CB & D \end{pmatrix}.
$$

So each row in $Y_{k|k+i-1}$ is a combination of the rows of X_k and $U_{k|k+i-1}$. Assuming that these rows are independent for the large j we did choose, which is natural in this context, we have that

$$
\text{rank } H_{k|k+i-1} = mi + n, \quad (j \geq mi + n), \tag{3.2}
$$

and also

$$
\text{rank } \begin{pmatrix} H_{1|i} \\ H_{i+1|2i} \end{pmatrix} = 2mi + n, \quad (j \geq 2mi + n). \tag{3.3}
$$

Observe that from (3.2) and (3.3) we can deduce the value of n. In fact something stronger holds, as stated in the following theorem.

For a matrix M we denote the row space, i.e., the span of the row vectors, by row M.

Theorem 3.5.1 *Assume that* (3.2) *and* (3.3) *hold true. Then*

$$\text{row } X_{i+1} = \text{row } H_{1|i} \cap \text{row } H_{i+1|2i}. \tag{3.4}$$

Proof   Since

$$\dim \text{row} \begin{pmatrix} H_{1|i} \\ H_{i+1|2i} \end{pmatrix} = \dim \text{row } H_{1|i} + \dim \text{row } H_{i+1|2i} - \dim \left(\text{row } H_{1|i} \cap \text{row } H_{i+1|2i} \right),$$

we conclude from (3.2) and (3.3) that $\dim \left(\text{row } H_{1|i} \cap \text{row } H_{i+1|2i} \right) = n$. Since we assume that Γ has full rank, there exists a matrix Γ^+ such that $\Gamma^+ \Gamma = I_n$. Therefore (3.1) gives that $X_{i+1} = \Gamma^{-1} Y_{i+1|2i} - \Gamma^+ H_T U_{i+1|2i}$ and thus row $X_{i+1} \subset$ row $H_{i+1|2i}$. Similarly we find that $X_1 \subset$ row $H_{1|i}$. Note that

$$X_{i+1} = A^i X_1 + \left(A^{i-1} B \ A^{i-2} B \ \cdots \ B \right) U_{1|i}.$$

Therefore also row $X_{i+1} \subset$ row $H_{1|i}$. Now use that both row X_{i+1} and row $H_{1|i} \cap$ row $H_{i+1|2i}$ have dimension n, to conclude the equality (3.4).   □

We are now ready to describe the algorithm to determine a system (A, B, C, D) that with the given input sequence $(u_k)_k$ generates the given output sequence $(y_k)_k$.

The first step is to find a basis for the space row $H_{1|i} \cap$ row $H_{i+1|2i}$. We know that such a basis consists of n row vectors. A way to compute such a basis will be described below. Here we use these n row vectors as rows for a matrix X'_{i+1}. The row space of X'_{i+1} is then equal to the row space of X_{i+1}. So $QX'_{i+1} = X_{i+1}$, where Q is an unknown invertible matrix. Also we know that the rows of X'_{i+1} are combinations of the rows of $H_{i+1|2i}$. So we have that $X'_{i+1} = T H_{i+1|2i}$ for some $n \times (\ell + m)i$ matrix T. Here T represents the way the basis is constructed from all rows. We put $X'_i = T H_{i|2i-1}$. All but one column (the first) of X'_i corresponds with a column of X'_{i+1}. For these columns we know that multiplying with Q gives the corresponding column of X_i. Since we took j large it follows that also $QX'_i = X_i$.

If we would have possession of A and B we would have $X_{i+1} = AX_i + BU_{i|i}$ and thus $X'_{i+1} = Q^{-1} A Q X'_i + Q^{-1} B U_{i|i}$. But we know neither Q nor A and B. On the other hand knowing $Q^{-1} A Q$ and $Q^{-1} B$ is good enough. This brings us at the second step. Solve A' and B' from the equation

$$X'_{i+1} = A' X'_i + B' U_{i|i}.$$

With A', B' and $U_{i|i}$ we construct the corresponding states $(\tilde{x}_k)_k$ and put

$$\tilde{X}_i = \left(\tilde{x}_i \cdots \tilde{x}_{i+j-1} \right).$$

Then solve C' and D' from the equation

$$Y_{i|i} = C'\tilde{X}_i + D'U_{i|i}.$$

This way we have found system matrices (A', B', C', D') such that the system

$$x_{k+1} = A'x_k + B'u_k,$$
$$y_k = C'x_k + D'u_k,$$

generates the given output sequence $(y_k)_k$ from the given input sequence $(u_k)_k$ with $x_0 = 0$.

Finally it remains to describe a method to actually obtain a basis for the space row $H_{1|i}$ $\cap$ row $H_{i+1|2i}$. First we use the singular value decomposition

$$\begin{pmatrix} H_{1|i} \\ H_{i+1|2i} \end{pmatrix} = \begin{pmatrix} U_{11} & U_{12} \\ U_{21} & U_{22} \end{pmatrix} \begin{pmatrix} S_{11} & 0 \\ 0 & 0 \end{pmatrix} V^T = USV^T \tag{3.5}$$

Here the sizes of the matrices are

$$U_{11} : (mi + \ell i) \times (2mi + n), \quad U_{12} : (mi + \ell i) \times (2\ell i - n)$$
$$U_{21} : (mi + \ell i) \times (2mi + n), \quad U_{22} : (mi + \ell i) \times (2\ell i - n)$$
$$S_{11} : (2mi + n) \times (2mi + n).$$

By applying U^T to the left of (3.5) and considering the last $2\ell i - n$ rows, we get that

$$U_{12}^T H_{1|i} + U_{22}^T H_{i+1|2i} = 0.$$

So the $(2\ell i - n)$ rows of $U_{22}^T H_{i+1|2i}$ are in row $H_{1|i}$ and in row $H_{i+1|2i}$ and span the intersection of these row spaces. Therefore a basis of row $\left(U_{22}^T H_{i+1|2i} \right)$ is also a basis of X_{i+1}. To find a basis of row $\left(U_{22}^T H_{i+1|2i} \right)$ we once again use the singular value decomposition and write

$$U_{22}^T H_{i+1|2i} = U'S'(V')^T,$$

with U' of size $(2\ell i - n) \times (2\ell i - n)$, S' of size $(2\ell i - n) \times j$, and $(V')^T$ of size $j \times j$. However, S' has only n nonzero singular values on its diagonal. Write

$$S' = \begin{pmatrix} S'_{11} & 0 \\ 0 & 0 \end{pmatrix}, \qquad (V')^T = \begin{pmatrix} V'_{11} & V'_{12} \\ V'_{21} & V'_{22} \end{pmatrix},$$

where S'_{11} is $n \times n$, and $\begin{pmatrix} V'_{11} & V'_{12} \end{pmatrix}$ is $n \times j$. The rows of $\begin{pmatrix} V'_{11} & V'_{12} \end{pmatrix}$ form an orthogonal basis of row $\left(U^T_{22} H_{i+1|2i} \right)$ and hence of row $H_{1|i} \cap$ row $H_{i+1|2i}$. So we choose the desired X'_{i+1} by putting $X'_{i+1} = \begin{pmatrix} V'_{11} & V'_{12} \end{pmatrix}$. Finally we also need the matrix X'_i. To that end we notice that

$$\begin{pmatrix} V'_{11} & V'_{12} \end{pmatrix} = (S'_{11})^{-1} \begin{pmatrix} U'_{11} \\ U'_{21} \end{pmatrix}^T U^T_{22} H_{i+1|2i} = T H_{i+1|2i},$$

where $T = (S'_{11})^{-1} \begin{pmatrix} U'_{11} \\ U'_{21} \end{pmatrix}^T U^T_{22}$, and hence we choose

$$X'_i = T H_{i|2i-1} = (S'_{11})^{-1} \left((U'_{11})^T \ (U'_{21})^T \right) U^T_{22} H_{i|2i-1}.$$

3.6   An Example

Let

$$A_0 = \begin{pmatrix} -1 & 1 \\ 0 & 1 \end{pmatrix}, \quad B_0 = \begin{pmatrix} 0 \\ 1 \end{pmatrix}, \quad C_0 = \begin{pmatrix} 1 & 0 \end{pmatrix}, \quad D_0 = 1,$$

and consider the system

$$x_{k+1} = A_0 x_k + B_0 u_k, \quad y_k = C_0 x_k + D_0 u_k,$$

Then, with $x_0 = 0$ and with input sequence

$$u = \begin{pmatrix} 1 & 0 & -1 & 1 & 0 & -2 & 1 & 0 & -3 & 1 & 0 & -4 & 1 & 0 & -5 \cdots \end{pmatrix}$$

we get the outputs

$$y = \begin{pmatrix} 1 & 0 & 0 & 1 & 0 & -1 & 1 & -1 & -2 & 0 & -2 & -4 & -1 & -4 & -6 \cdots \end{pmatrix}.$$

We will use the subspace algorithm to determine a system from the u_k and y_k that generates the output sequence $(y_k)_{k=1}^{\infty}$ from the input sequence $(u_k)_{k=1}^{\infty}$.

We choose numbers i and j with i not to small (larger than the estimated state space dimension) and j much larger than i. So we take $i = 3$ and $j = 10$. Note that $m = \ell = 1$ for the system we have under consideration. From the given data we construct $2i \times j = 6 \times 10$—matrices

$$
H_{1|3} = \begin{pmatrix} u_1 & u_2 & \cdots & u_{10} \\ u_2 & u_3 & \cdots & u_{11} \\ u_3 & u_4 & \cdots & u12 \\ y_1 & y_2 & \cdots & y_{10} \\ y_2 & y_3 & \cdots & y_{11} \\ y_3 & y_4 & \cdots & y_{12} \end{pmatrix}, \quad H_{4|6} = \begin{pmatrix} u_4 & u_5 & \cdots & u_{13} \\ u_5 & u_6 & \cdots & u_{14} \\ u_6 & u_7 & \cdots & u_{15} \\ y_4 & y_5 & \cdots & y_{13} \\ y_5 & y_6 & \cdots & y_{14} \\ y_6 & y_7 & \cdots & y_{15} \end{pmatrix}.
$$

Then

$$
H_{1|3} = \begin{pmatrix}
1 & 0 & -1 & 1 & 0 & -2 & 1 & 0 & -3 & 1 \\
0 & -1 & 1 & 0 & -2 & 1 & 0 & -3 & 1 & 0 \\
-1 & 1 & 0 & -2 & 1 & 0 & -3 & 1 & 0 & -4 \\
1 & 0 & 0 & 1 & 0 & -1 & 1 & -1 & -2 & 0 \\
0 & 0 & 1 & 0 & -1 & 1 & -1 & -2 & 0 & -2 \\
0 & 1 & 0 & -1 & 1 & -1 & -2 & 0 & -2 & -4
\end{pmatrix}
$$

$$
H_{4|6} = \begin{pmatrix}
1 & 0 & -2 & 1 & 0 & -3 & 1 & 0 & -4 & 1 \\
0 & -2 & 1 & 0 & -3 & 1 & 0 & -4 & 1 & 0 \\
-2 & 1 & 0 & -3 & 1 & 0 & -4 & 1 & 0 & -5 \\
1 & 0 & -1 & 1 & -1 & -2 & 0 & -2 & -4 & -1 \\
0 & -1 & 1 & -1 & -2 & 0 & -2 & -4 & -1 & -4 \\
-1 & 1 & -1 & -2 & 0 & -2 & -4 & -1 & -4 & -6
\end{pmatrix}
$$

We determine a basis for the intersection of the row spaces of $H_{1|3}$ and $H_{4|6}$. Although it is not so straightforward, one can check that the rows of the matrix X'_4 given by

$$
X'_4 = \begin{pmatrix}
0 & 1 & 0 & -1 & 1 & -1 & -2 & 0 & -2 & -4 \\
0 & 0 & 1 & 0 & -1 & 1 & -1 & -2 & 0 & -2
\end{pmatrix}
$$

are indeed a basis for the intersection of the row spaces of $H_{1|3}$ and $H_{4|6}$. Indeed, from the theory we already know that the dimension of the intersection of these row spaces is $2(mi + n) - (2mi + n) = n$, and by inspection we see that the ranks of H_{13} and H_{14} are both 5, while by computation one can see that the rank of $\begin{pmatrix} H_{13} \\ H_{24} \end{pmatrix}$ is 8. So $n = 2$.

To be more precise, we have that

$$X'_4 = \begin{pmatrix} 0\,0\,1\,1\,0\,0 \\ 0\,0\,0\,0\,1\,0 \end{pmatrix} H_{13}$$

and

$$X'_4 = \begin{pmatrix} 0 & -1\,0\,0\,1\,0 \\ -1 & 0\ \ 0\,1\,0\,0 \end{pmatrix} H_{46}.$$

This also shows that we can take

$$T = \begin{pmatrix} 0 & -1\,0\,0\,1\,0 \\ -1 & 0\ \ 0\,1\,0\,0 \end{pmatrix}.$$

Next we use T to determine the X'_3, the 'states' one unit in time earlier, by $X'_3 = T H_{35}$, where

$$H_{35} = \begin{pmatrix} -1 & 1 & 0 & -2 & 1 & 0 & -3 & 1 & 0 & 4 \\ 1 & 0 & -2 & 1 & 0 & -3 & 1 & 0 & -4 & 1 \\ 0 & -2 & 1 & 0 & -3 & 1 & 0 & -4 & 1 & 0 \\ 0 & 1 & 0 & -1 & 1 & -1 & -2 & 0 & -2 & -4 \\ 1 & 0 & -1 & 1 & -1 & -2 & 0 & -2 & -4 & -1 \\ 0 & -1 & 1 & -1 & -2 & 0 & -2 & -4 & -1 & -4 \end{pmatrix}$$

Then

$$X'_3 = \begin{pmatrix} 0\,0\,1\,0 & -1 & 1 & -1 & -2 & 0 & -2 \\ 1\,0\,0\,1 & 0 & -1 & 1 & -1 & -2 & 0 \end{pmatrix}$$

We next determine A and B from the equation

$$X'_4 = AX'_3 + B \left(u_3\ u_4\ \cdots\ u_{12} \right) = \begin{pmatrix} A & B \end{pmatrix} \begin{pmatrix} X'_3 \\ U_{3|3} \end{pmatrix}$$

So

$$\begin{pmatrix} X'_3 \\ U_{3|3} \end{pmatrix} = \begin{pmatrix} 0 & 0\,1 & 0 & -1 & 1 & -1 & -2 & 0 & -2 \\ 1 & 0\,0 & 1 & 0 & -1 & 1 & -1 & -2 & 0 \\ -1 & 1\,0 & -2 & 1 & 0 & -3 & 1 & 0 & -4 \end{pmatrix}.$$

Solving for $\begin{pmatrix} A & B \end{pmatrix}$ from

$$\begin{pmatrix} A & B \end{pmatrix} \begin{pmatrix} X_3' \\ U_{3|3} \end{pmatrix} = X_4',$$

one finds

$$A = \begin{pmatrix} 0 & 1 \\ 1 & 0 \end{pmatrix}, \qquad B = \begin{pmatrix} 1 \\ 0 \end{pmatrix}.$$

With A and B we generate, starting at $\tilde{x}_0 = 0$ a new set of states $\tilde{x}_k$

$$\tilde{X} = \begin{pmatrix} 0 & 1 & 0 & 0 & 1 & 0 & -1 & 1 & -1 & -2 & 0 & -2 & -4 & -1 & -4 & -6 \cdots \\ 0 & 0 & 1 & 0 & 0 & 1 & 0 & -1 & 1 & -1 & -2 & 0 & -2 & -4 & -1 & -4 \cdots \end{pmatrix}$$

Then $\tilde{X}_3$ is the matrix composed of the columns 3 to 12 of this matrix:

$$\tilde{X}_3 = \begin{pmatrix} 0 & 0 & 1 & 0 & -1 & 1 & -1 & -2 & 0 & -2 \\ 1 & 0 & 0 & 1 & 0 & -1 & 1 & -1 & -2 & 0 \end{pmatrix}$$

and solving C and D from $Y_{3|3} = C\tilde{X}_3 + DU_{3|3}$ we obtain

$$C = \begin{pmatrix} 0 & 1 \end{pmatrix}, \qquad D = 1$$

as indeed

$$Y_{3|3} = \begin{pmatrix} 0 & 1 & 0 & -1 & 1 & -1 & -2 & 0 & -2 & -4 \end{pmatrix}$$

$$= \begin{pmatrix} 0 & 1 & 1 \end{pmatrix} \begin{pmatrix} 0 & 0 & 1 & 0 & -1 & 1 & -1 & -2 & 0 & -2 \\ 1 & 0 & 0 & 1 & 0 & -1 & 1 & -1 & -2 & 0 \\ -1 & 1 & 0 & -2 & 1 & 0 & -3 & 1 & 0 & -4 \end{pmatrix}.$$

The system (A, B, C, D) indeed generates the outputs y_k from the inputs u_k. Notice that the system (A, B, C, D) is similar to the system (A_0, B_0, C_0, D_0), which we started with. To be precise $A_0 = QAQ^{-1}$, $QB = B_0$ and $CQ^{-1} = C_0$ with $Q = \begin{pmatrix} 0 & 1 \\ 1 & 1 \end{pmatrix}$.

Stability

4

Input-output systems are applied in control, where the inputs are chosen in such a way that the system shows satisfactory performance. Stability is an important objective, that is, disturbances have a limited effect on the system. Systems can be stabilized by feedback, where past performance is used to choose the input variables.

4.1 Internal Stability

Stated in general terms, a system is stable if perturbations have no long lasting effects. That is, if a system at rest is brought out of equilibrium, then the dynamics tends to bring the system back to its original position. If a system is not stable, then we may wish to make it stable by applying an appropriate control input to the system. In this chapter we consider these questions for linear systems. As a first step we consider the stability of the state vector when no control is applied. The system is then given by the equation

$$x(t + 1) = Ax(t), \tag{4.1}$$

where A is an $n \times n$ matrix with real entries. Clearly, the zero vector is an equilibrium, i.e. the function $x(t) \equiv 0$ is a constant solution, and the question is whether the state vector tends to zero when started at $x(0) = x_0 \neq 0$.

Definition 4.1.1 The system (4.1) is called *asymptotically stable* if $x(t) \to 0$ for $t \to \infty$ for every initial value $x(0) = x_0 \in \mathbb{R}^n$.

We describe two methods to check the stability of (4.1), one in terms of eigenvalues and the other in terms of linear matrix inequalities.

© The Author(s), under exclusive license to Springer Nature Switzerland AG 2021
C. Heij et al., *Introduction to Mathematical Systems Theory*,
https://doi.org/10.1007/978-3-030-59654-5_4

Theorem 4.1.2 *The system* (4.1) *is asymptotically stable if and only if A has all its eigenvalues in the open unit disc.*

Proof   For simplicity we prove this result only under the simplifying assumption that A is diagonalizable, The general case requires the use of the Jordan canonical form of A. Readers that are familiar with the Jordan canonical form may adjust the following argument to prove the general case.

Assume $A = S\text{diag}\,(\lambda_1, \ldots, \lambda_n)S^{-1}$, then $A^k = S\text{diag}\,(\lambda_1^k, \ldots, \lambda_n^k)S^{-1}$. If all $|\lambda_i| < 1$ then $\lambda_i^t \to 0$ as $t \to \infty$ for all i. Therefore $x(t) = A^t x_0 \to 0$ for $t \to \infty$ for every initial vector $x(0) = x_0$. Conversely, suppose $|\lambda_i| \geq 1$ for some i, say for $i = 1$. Let $x(0) = x_0$ be a (complex) eigenvector of A corresponding to λ_1, then $x(t) = A^k x_0 = \lambda_1^t x_0$, and $\|x(t)\| = |\lambda_1|^t \|x_0\|$. This does not tend to zero, so that (4.1) is not asymptotically stable.

$\square$

The matrix A is called *stable* if all its eigenvalues are in the open unit disc. The next result is a test on the stability of a matrix in terms of positive definite matrices.

Theorem 4.1.3 *An $n \times n$ matrix A is stable if and only if there is a positive definite matrix P such that $P - A^T P A$ is positive definite.*

This is a corollary of the following result; it will be proved after the next result.

Theorem 4.1.4 *Let (A, C) be observable, then A is stable if and only if there is a positive definite solution of the equation*

$$P - A^T P A = C^T C. \tag{4.2}$$

In that case P is unique and is given by

$$P = \sum_{j=0}^{\infty} (A^T)^j C^T C A^j. \tag{4.3}$$

Proof   First suppose that A is stable. For simplicity we assume that A is diagonalizable, the general case uses the Jordan canonical form again. So let $A = S\text{diag}\left(\lambda_1 \ldots \lambda_n\right) S^{-1}$ with $m = \max_{1 \leq i \leq n} |\lambda_i| < 1$. We first show that (4.3) is a convergent series. Now $A^j = S\text{diag}\left(\lambda_1^j \ldots \lambda_n^j\right) S^{-1}$, so that (with the induced matrix norm) $\|A^j\| \leq \|S\| \cdot \|S^{-1}\| m^j$. Therefore

$$\|(A^T)^j C^T C A^j\| \leq \|(A^T)^j\| \|C^T C\| \|A^j\| \leq \|S^T\| \|(S^{-1})^T\| \cdot \|S\| \|S^{-1}\| \|C^T C\| \cdot m^{2j},$$

so that $\|(A^T)^j C^T C A^j\| \leq c_0 m^{2j}$ for some constant c_0. Hence the series (4.3) converges. It is easy to see that P as defined in (4.3) satisfies (4.2), and as $(A^T)^j C^T C A^j$ is positive semidefinite for all j it follows that also P is positive semidefinite. It remains to show that P is nonsingular. Suppose $Px = 0$, then (4.3) implies that $(A^T)^j C^T C A^j x = 0$ and hence $\langle (A^T)^j C^T C A^j x, x \rangle = \langle C A^j x, C A^j x \rangle = \|C A^j x\|^2 = 0$. So $C A^j x = 0$ for $j \geq 0$ and as (A, C) is observable this implies $x = 0$. The solution of (4.2) is also unique in this case. Indeed, let Q be a solution of (4.2), let P be given by (4.3). Then

$$PQ^{-1} = \sum_{j=0}^{\infty}(A^T)^j C^T C A^j Q^{-1}$$

$$= \sum_{j=0}^{\infty}(A^T)^j Q A^j Q^{-1} - \sum_{j=0}^{\infty}(A^T)^j A^T Q A A^j Q^{-1}$$

$$= \sum_{j=0}^{\infty}(A^T)^j Q A^j Q^{-1} - \sum_{j=1}^{\infty}(A^T)^j Q A^j Q^{-1} = QQ^{-1} = I,$$

so that $P = Q$.

Conversely, suppose there is a solution P of (4.2). Then we have to show that A is stable. Let λ be an eigenvalue of A, so that $Ax = \lambda x$ with $x \neq 0$. Then

$$\|Cx\|^2 = \langle C^T Cx, x \rangle = \langle (P - A^T PA)x, x \rangle =$$

$$= \langle Px, x \rangle - \langle PAx, Ax \rangle = (1 - |\lambda|^2)\langle Px, x \rangle.$$

As P is positive definite, either $|\lambda| < 1$ or $|\lambda| = 1$ with $Cx = 0$. However, $Cx = 0$ and $Ax = \lambda x$, $x \neq 0$ is impossible as (A, C) is observable, see Theorem 3.2.2. Thus $|\lambda| < 1$, so that A is stable. $\qquad\square$

Proof *(of Theorem 4.1.3)* Suppose that there exists a $P > 0$ such that $P - A^T PA > 0$. Put $V = P - A^T PA$. This matrix is symmetric and positive definite. Hence there is a unitary matrix U and a positive diagonal matrix Λ such that $V = U \Lambda U^*$. Put $C = V^{\frac{1}{2}} = U \Lambda^{\frac{1}{2}} U^*$. Then $P - A^T PA = C^T C$. Since $C = V^{\frac{1}{2}}$ is positive definite, it is invertible and hence (A, C) is observable. From Theorem 4.1.4 it now follows that A is stable.

The converse is immediate from Theorem 4.1.4 by taking $C = I$. $\qquad\square$

The dual version of Theorem 4.1.4 is the following result.

Theorem 4.1.5 *Let (A, B) be controllable, then A is stable if and only if there is a positive definite solution of the equation*

$$Q - AQA^T = BB^T. \tag{4.4}$$

In that case Q is unique and is given by

$$Q = \sum_{j=0}^{\infty} A^j B B^T (A^T)^j. \tag{4.5}$$

Equations (4.2) and (4.4) are called *Stein equations* (or *discrete Lyapunov equations*). Their solutions P and Q given by (4.3) and (4.5) are called, respectively, the *observability Grammian* and *controllability Grammian* of (A, B, C).

As an illustration we consider the stability of the second order difference equation

$$y(t) = ay(t-1) + by(t-2).$$

Introduce as state $x(t) = \begin{pmatrix} y(t-1) \\ y(t-2) \end{pmatrix}$, then the equation can be written as

$$x(t+1) = \begin{pmatrix} a & b \\ 1 & 0 \end{pmatrix} x(t)$$

$$y(t) = \begin{pmatrix} a & b \end{pmatrix} x(t).$$

The equilibrium solution $y(t) \equiv 0$ is asymptotically stable if and only if the equilibrium state $x(t) \equiv 0$ is asymptotically stable. The eigenvalues of the state transition matrix are given by

$$\lambda_{1,2} = \tfrac{1}{2}a \pm \tfrac{1}{2}\sqrt{a^2 + 4b} \qquad \text{if } a^2 > -4b,$$

$$\lambda_{1,2} = \tfrac{1}{2}a \pm \tfrac{i}{2}\sqrt{-a^2 - 4b} \qquad \text{if } a^2 \leq -4b.$$

We consider three cases: (i) $a^2 + 4b > 0$, $a > 0$; (ii) $a^2 + 4b > 0$, $a < 0$; (iii) $a^2 + 4b \leq 0$. In case (i) $|\lambda_{1,2}| < 1$ if and only if $\tfrac{1}{2}a + \tfrac{1}{2}\sqrt{a^2 + 4b} < 1$, so that $0 \leq \tfrac{1}{2}\sqrt{a^2 + 4b} < 1 - \tfrac{1}{2}a$, that is, $0 < a < 2$ and $b < 1 - a$. In case (ii) $|\lambda_{1,2}| < 1$ if and only if $\tfrac{1}{2}a - \tfrac{1}{2}\sqrt{a^2 + 4b} > -1$, so that $0 \geq -\tfrac{1}{2}\sqrt{a^2 + 4b} > -1 - \tfrac{1}{2}a$, that is, $-2 < a < 0$ and $b < 1 + a$. In case (iii) $|\lambda_1| = |\lambda_2|$, so $|\lambda_{1,2}| < 1$ if and only if $(\tfrac{1}{2}a)^2 + (-b - \tfrac{1}{4}a^2) = -b < 1$. Combining these results, the system is stable if and only if (a, b) lies in the triangle in the following Fig. 4.1.

Example 4.1.1 In Sect. 1.4 we considered a simple model for the macro-economic business cycle. Taking the variables in deviation from their equilibrium values corresponding

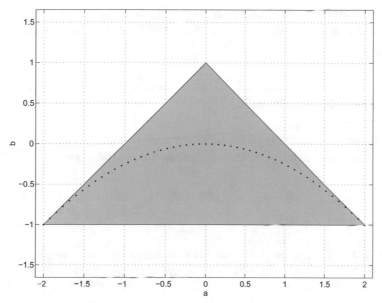

Fig. 4.1 The stability region for $y(t) = ay(t-1) + by(t-2)$

to a given level of government spending, the model is described by

$$c(t) = \beta y(t-1),$$

$$i(t) = \delta(c(t) - c(t-1)),$$

$$y(t) = c(t) + i(t) + g(t).$$

Here c denotes consumption, y national income, i investment and g government expenditures. Now assume that government spending is constant for $t \geq 0$, so that $g(t) = 0$ for $t \geq 0$. The economy is stable if the other variables tend to their equilibrium values corresponding to this level of government spending, that is, if $c(t)$, $y(t)$ and $i(t)$ all tend to zero if $t \to \infty$. Clearly it is necessary and sufficient that $y(t) \to 0$ for $t \to \infty$.

For $t \geq 0$ the dynamics of national income is described by (1.33) with $g(t) \equiv 0$, that is,

$$y(t) = \beta(1+\delta)y(t-1) - \beta\delta y(t-2)$$

This is a second order difference equation with coefficients $a = \beta(1+\delta)$ and $b = -\beta\delta$. From an economic point of view, the restrictions $\beta > 0$ and $\delta > 0$ are reasonable. This implies that $a > 0$ and $b < 0$. From the above result for general second order difference equations it follows that this system is asymptotically stable if and only if $b > -1$ and $a + b < 1$, that is, $\beta\delta < 1$ and $\beta < 1$. The last restriction is plausible for economic reasons, while the first restriction means that investors should not react too strongly to increased consumption. Finally, the path towards equilibrium will show oscillations if the characteristic roots of the equation are non-real, that is, if $a^2 + 4b < 0$ or, equivalently, $\beta < \frac{4\delta}{(1+\delta)^2}$.

4.2   Input-Output Stability

So far we considered stability of the state of a system. This internal stability is closely connected to external stability, which is defined as follows. We consider the system

$$
\begin{cases}
x(t+1) & = Ax(t) + Bu(t), \quad x(0) = 0, \\
y(t) & = Cx(t) + Du(t).
\end{cases}
\tag{4.6}
$$

Definition 4.2.1 The system (4.6) is called *externally stable* (or *bounded-input, bounded-output stable*) if for each $M > 0$ there exists $N > 0$ such that $\|u(t)\| \le M$ for $t \ge 0$ implies $\|y(t)\| \le N$ for $t \ge 0$.

The next result gives a criterion in terms of the impulse response of the system.

Theorem 4.2.2 (i) *The system* (4.6) *is externally stable if and only if*

$$
\sum_{j=0}^{\infty} \|G(j)\| < \infty,
$$

where $G(j) = CA^{j-1}B$ *is the impulse response of* (4.6) *and* $G(0) = D$.
(ii) *If A is a stable matrix, then the system is externally stable.*

Proof (i) Suppose that $\sum_{j=0}^{\infty} \|G(j)\| < \infty$ and let u be an input sequence with $\|u(t)\| \le M$ for all $t \ge 0$. Then

$$
\|y(t)\| = \left\| \sum_{j=0}^{t} G(t-j)u(j) \right\| \le \sum_{j=0}^{t} \|G(t-j)\| \|u(j)\| \le M \cdot \sum_{j=0}^{\infty} \|G(j)\|.
$$

So we can take $N = M \cdot \sum_{j=0}^{\infty} \|G(j)\|$.

The converse we will prove first for single input, single output systems. Assume that the system is externally stable. Let N be such that each input u with $|u(t)| \le 1$ for all t

gives an output y with $|y(t)| \leq N$. Now, for fixed j, define the input sequence v by

$$
v(t) = \begin{cases} \dfrac{|G(j-t)|}{G(j-t)} & \text{for all } 0 \leq t \leq j, \ G(j-t) \neq 0 \\ 0 & \text{for all other } t \geq 0. \end{cases}
$$

As $|v(t)| \leq 1$ for all t, the corresponding output satisfies $|y(t)| \leq N$ for all t and in particular,

$$
|y(j)| = \left| \sum_{t=0}^{j} G(j-t)v(t) \right| = \sum_{t=0}^{j} |G(j-t)| = \sum_{t=0}^{j} |G(t)| \leq N.
$$

As this holds for all j, it follows that $\sum_{t=0}^{\infty} |G(t)| \leq N < \infty$.

For the general case we assume that the system is externally stable. To fix notation, the input space is $\mathbb{R}^m$, the output space is $\mathbb{R}^p$. The single-input, single-output case one obtains by using as input the l'th coordinate of $u(t)$ and as output the k'th coordinate of $y(t)$ is also externally stable. Its impulse respons matrices are the 1×1 matrices $G(j)_{k,l}$. Therefore $\sum_{j=0}^{\infty} |G(j)_{k,l}| < \infty$. Now we use a well-known estimate of the norm of a matrix in terms of the absolute values of its entries: $\|G(j)\| \leq \sum_{k,l=1}^{p,m} |G(j)_{k,l}|$. So $\sum_{j=0}^{\infty} \|G(j)\| \leq \sum_{j=0}^{\infty} \sum_{k,l=1}^{p,m} |G(j)_{k,l}| = \sum_{k,l=1}^{p,m} \sum_{j=0}^{\infty} |G(j)_{k,l}| < \infty$.

(ii) We show this only for the case when A is diagonalizable, so that $S^{-1}AS = \text{diag}(\lambda_1, \ldots, \lambda_n)$. As before, the general case requires the use of the Jordan canonical form. Let the eigenvalues of A be ordered such that $|\lambda_1| \geq |\lambda_j|$ for $j = 1, \ldots, n$. Then $\|G(j)\| \leq \|C\| \cdot \|B\| \cdot \|S\| \cdot \|S^{-1}\| \cdot |\lambda_1|^{j-1} = c_0 |\lambda_1|^{j-1}$ for some constant c_0. If A is stable, then $|\lambda_1| < 1$, so that $\sum_{j=1}^{\infty} \|G(j)\| \leq c_0 \cdot \sum_{j=0}^{\infty} |\lambda_1|^j = \frac{c_0}{1-|\lambda_1|} < \infty$. □

Theorem 4.2.3 *Let (4.6) be a minimal realization. Then (4.6) is externally stable if and only if A is a stable matrix.*

Proof Given the results in Theorem 4.2.2, it remains to prove that stability of the matrix A follows from $\sum_{j=0}^{\infty} \|CA^j B\| < \infty$. Let λ be an eigenvalue of A, and let $x \neq 0$ be a corresponding eigenvector. As (A, C) is observable, we have $Cx \neq 0$. As (A, B) is controllable, there exist $T \geq 1$ and an input sequence $\{u(t)\}_{t=0}^{T-1}$ such that the state $x(T)$ at time T resulting from this input sequence is equal to x, that is, $\sum_{j=0}^{T-1} A^{T-1-j} Bu(j) = x$. Let this input sequence be continued with zero inputs from time T onwards, that is

$u(t) = 0$, $t \geq T$. The corresponding output is given by $y(t) = \sum_{j=0}^{t-1} CA^{t-j-1} Bu(j)$, and for $t \geq T$ this equals

$$y(t) = \sum_{j=0}^{T-1} CA^{t-j-1} Bu(j) = CA^{t-T} \sum_{j=0}^{T-1} A^{T-j-1} Bu(j) = \qquad (4.7)$$

$$= CA^{t-T}x = \lambda^{t-T} Cx.$$

Using the first equality in (4.7) and defining $M = \max_{0 \leq j \leq T-1} \|u(j)\|$, it follows that

$$\|y(t)\| \leq \sum_{j=0}^{T-1} \|CA^{t-j-1}B\| \|u(j)\| \leq M \sum_{j=0}^{T-1} \|CA^{t-j-1}B\| =$$

$$= M \sum_{i=t-T}^{t-1} \|CA^i B\| \to 0$$

for $t \to \infty$, because $\sum_{i=0}^{\infty} \|CA^i B\| < \infty$. As also $\|y(t)\| = |\lambda|^{t-T} \|Cx\|$ with $Cx \neq 0$, this can only converge to zero for $t \to \infty$ if $|\lambda| < 1$. This shows that A is a stable matrix. $\qquad \square$

Example 4.2.1 We consider the model for the demand process in Example 1.1.3, given by

$$\begin{pmatrix} T(t+1) \\ B(t+1) \\ S(t+1) \\ S(t) \\ S(t-1) \end{pmatrix} = \begin{pmatrix} 1 & 1 & 0 & 0 & 0 \\ 0 & 1 & 0 & 0 & 0 \\ 0 & 0 & -1 & -1 & -1 \\ 0 & 0 & 1 & 0 & 0 \\ 0 & 0 & 0 & 1 & 0 \end{pmatrix} \begin{pmatrix} T(t) \\ B(t) \\ S(t) \\ S(t-1) \\ S(t-2) \end{pmatrix} + \begin{pmatrix} \eta(t+1) \\ \xi(t+1) \\ \omega(t+1) \\ 0 \\ 0 \end{pmatrix}$$

$$D(t) = T(t) + S(t) + \varepsilon(t).$$

Here D is the observed demand, T is the trend term with slope B, S is a seasonal term, and η, ξ, ω and ε are noise terms. Taking D as output and $\begin{pmatrix} \varepsilon & \eta & \xi & \omega \end{pmatrix}^T$ as input variables, it is easily checked that this realization is controllable and observable, hence minimal. If all noise terms are zero, then $D(t) = 0$ is an equilibrium. However, this equilibrium is not asymptotically stable because the state transition matrix does not have all its eigenvalues within the unit disc. In fact, the characteristic polynomial is given by $(\lambda - 1)^2(\lambda + 1)(\lambda + i)(\lambda - i)$, so that all the eigenvalues are exactly on the unit circle.

4.3   Stabilization by State Feedback

In control applications, one of the basic objectives is to construct stable systems. If a system is not stable by itself, the question arises whether it can be stabilized by choosing the control inputs appropriately. This is called the stabilization problem. More in particular, we wish to construct a control law such that the system is brought to rest from any given initial position. The idea is that the system may be excited by external disturbances, and that the control inputs should eliminate these effects.

We consider the input-output system described by

$$\begin{cases} x(t+1) & = Ax(t) + Bu(t), \\ y(t) & = Cx(t). \end{cases} \tag{4.8}$$

The fundamental idea of feedback control is to use the past information on inputs and outputs to choose the current value of the input. As the state variable summarizes all past information that is relevant for the future, this suggests to consider so-called *static state feedback controllers* of the form

$$u(t) = Fx(t), \tag{4.9}$$

where F is an $m \times n$ matrix. In this section we assume that the state $x(t)$ is known at time t, so that this control can be implemented. In the next section we discuss controllers for the case that the state is not directly observed, so that it should be reconstructed from the observed inputs and outputs.

The *closed loop system* obtained by applying the control law (4.9) to the system (4.8) has state equation

$$x(t+1) = (A + BF)x(t), \qquad x(0) = x_0.$$

The stability of this system depends on the matrix $A + BF$, where A and B are given and F has to be constructed.

Theorem 4.3.1 *Let A be an $n \times n$ matrix and B an $n \times m$ matrix. The pair (A, B) is controllable if and only if for every monic polynomial $p(\lambda) = \lambda^n + p_{n-1}\lambda^{n-1} + \cdots + p_1\lambda + p_0$ there exists an $m \times n$ matrix F such that*

$$\det(\lambda I_n - (A + BF)) = p(\lambda).$$

This result is called the *pole placement theorem*. It shows that for controllable systems the feedback law (4.9) can achieve any desired level of stability. For example, by an appropriate choice of (4.9) the closed loop polynomial is equal to λ^n. This means that

$(A + BF)^n = 0$, that is, the system is back at equilibrium in finite time, after n time periods. Such a controller is called a *dead-beat controller*.

Proof   We prove this only for systems with a single input, so that $m = 1$.

First assume that (A, B) is controllable. We shall first show that in this case we may assume that in an appropriately chosen basis we have

$$A = \begin{pmatrix} 0 & 1 & & \\ & & \ddots & \\ & & & 1 \\ -a_0 & \cdots\cdots & -a_{n-1} \end{pmatrix}, \qquad B = \begin{pmatrix} 0 \\ \vdots \\ 0 \\ 1 \end{pmatrix}. \tag{4.10}$$

Indeed, let $\det(\lambda I - A) = a_0 + a_1\lambda + \cdots + a_{n-1}\lambda^{n-1} + \lambda^n = a(\lambda)$. Now we define vectors s_j for $j = 1, \ldots, n$ as follows: $s_1 = b$ and $s_{j+1} = As_j + a_{n-j}b$, and we define a matrix S by $S = \begin{pmatrix} s_n \cdots s_1 \end{pmatrix}$. Then by using the Cayley-Hamilton theorem $(0 = a_0 + a_1 A + \cdots + a_n A^n)$ one sees that

$$AS = S \begin{pmatrix} 0 & 1 & & \\ & & \ddots & \\ & & & 1 \\ -a_0 & \cdots\cdots & -a_{n-1} \end{pmatrix}, \qquad S^{-1}B = \begin{pmatrix} 0 \\ \vdots \\ 0 \\ 1 \end{pmatrix}.$$

So, we may as well assume that A and B are given by (4.10). Recall that $\det(\lambda I - A)$ is equal to $a(\lambda)$. Define $F = \begin{pmatrix} a_0 - p_0 & a_1 - p_1 & \cdots & a_{n-1} - p_{n-1} \end{pmatrix}$, then

$$A + BF = \begin{pmatrix} 0 & 1 & & \\ & & \ddots & \\ & & & 1 \\ -p_0 & \cdots\cdots & -p_{n-1} \end{pmatrix}.$$

Now note that, analogously to $\det(\lambda I - A) = a(\lambda)$ we have $\det(\lambda I - (A + BF)) = p(\lambda)$.

Conversely, if (A, B) is not controllable then by an appropriate choice of basis we may write

$$A = \begin{pmatrix} A_{11} & A_{12} \\ 0 & A_{22} \end{pmatrix}, \qquad B = \begin{pmatrix} B_1 \\ 0 \end{pmatrix}.$$

In this case the characteristic polynomial of $A + BF$ will always have roots at the eigenvalues of A_{22}. □

As minimal realizations are controllable, the pole placement theorem solves the state feedback stabilization problem for minimal systems. In practice it may also be of interest to consider non-minimal realizations and to investigate whether all states in a non-minimal realization are stable. For this purpose we use the concept of stabilizability.

Definition 4.3.2 The pair (A, B) is called *stabilizable* if there exists a matrix F such that $A + BF$ is stable. In this case any matrix F such that $A + BF$ is stable is called a *stabilizing feedback matrix*.

That (A, B) is stabilizable if and only if the system (4.8) can be stabilized by the static state feedback (4.9). The next result shows under which conditions stabilization is possible.

Theorem 4.3.3 *The pair (A, B) is stabilizable if and only if every (A, B)-uncontrollable eigenvalue of A lies in the open unit disc.*

Proof Let (A, B) be stabilizable and let λ be an uncontrollable eigenvalue of A, so that $x^T \left(A - \lambda I \ B \right) = 0$ for some $x \neq 0$. Then for every F we have $x^T (A + BF) = x^T A = \lambda x^T$, so that λ is an eigenvalue of $A + BF$. Stabilizability implies that $|\lambda| < 1$.

Conversely, if every uncontrollable eigenvalue of A lies within the open unit disc, then after basis transformation we can write

$$A = \begin{pmatrix} A_{11} & A_{12} \\ 0 & A_{22} \end{pmatrix}, \qquad B = \begin{pmatrix} B_1 \\ 0 \end{pmatrix},$$

where (A_{11}, B_1) is controllable and all eigenvalues of A_{22} are in the open unit disc. By Theorem 4.3.1 there exists a matrix F_1 such that $A_{11} + B_1 F_1$ is stable. Now define $F = \left(F_1 \ 0 \right)$, then

$$A + BF = \begin{pmatrix} A_{11} + B_1 F_1 & A_{12} \\ 0 & A_{22} \end{pmatrix},$$

is a stable matrix.   □

Example 4.3.1 In Sect. 1.4 we derived the following state space model

$$x(t+1) = \begin{pmatrix} \beta(1+\delta) & -\beta\delta \\ 1 & 0 \end{pmatrix} x(t) + \begin{pmatrix} 1 \\ 0 \end{pmatrix} g(t),$$

$$\begin{pmatrix} c(t) \\ i(t) \\ y(t) \end{pmatrix} = \begin{pmatrix} \beta & 0 \\ \beta\delta & -\beta\delta \\ \beta(1+\delta) & -\beta\delta \end{pmatrix} x(t) + \begin{pmatrix} 0 \\ 0 \\ 1 \end{pmatrix} g(t).$$

Here the state vector is defined by $x(t) = (y(t-1), y(t-2))^T$.

Assuming that $\delta > 0$ and $0 < \beta < 1$, we showed in Example 4.1.1 that this system is asymptotically stable if and only if $\beta\delta < 1$. Now suppose that $\beta\delta \geq 1$. As the system is controllable, it is certainly stabilizable. If $F = \begin{pmatrix} f_1 & f_2 \end{pmatrix}$, then the feedback law (4.9) is given by

$$g(t) = f_1 y(t-1) + f_2 y(t-2).$$

Combining this with (1.33), this gives the closed loop system

$$y(t) = (\beta(1+\delta) + f_1)y(t-1) + (-\beta\delta + f_2)y(t-2).$$

The conditions for stability are described by the triangle in Fig. 4.1, with $a = \beta(1+\delta) + f_1$ and $b = -\beta\delta + f_2$. In particular, if $\beta\delta \geq 1$, then it follows that the system cannot be stabilized with $f_2 = 0$, that is, the government cannot stabilize the system if it only considers the deviation from equilibrium occurring in the last year. It has to take into account also the year before. If, for example, we take $f_2 = \beta\delta$ and $f_1 = -\beta(1+\delta)$, then we obtain a dead beat controller that brings the economy back to equilibrium in two periods of time.

4.4   Stabilization by Output Feedback

In the foregoing we assumed that the state is observed, so that the feedback law (4.9) can be implemented. Now suppose that $x(t)$ in (4.6) is not available at time t. The control input can then be based on past inputs and outputs. Also, we assume that the controller is described by the system

$$\begin{cases} z(t+1) & = Mz(t) + Nu(t) + Ly(t), \\ u(t) & = Fz(t). \end{cases} \tag{4.11}$$

This is called a *dynamic compensator*. It is called a *stabilizing compensator* or *stabilizing dynamic feedback* if the closed loop system composed of (4.6) and (4.11) is stable, that is, if for every initial values $x(0)$ and $z(0)$ all signals $u(t)$, $y(t)$, $x(t)$ and $z(t)$ tend to zero for $t \to \infty$.

The closed loop system is described by the equations

$$
\begin{cases}
\begin{pmatrix} x(t+1) \\ z(t+1) \end{pmatrix} = \begin{pmatrix} A & BF \\ LC & M+(N+LD)F \end{pmatrix} \begin{pmatrix} x(t) \\ z(t) \end{pmatrix}, \\
\begin{pmatrix} y(t) \\ u(t) \end{pmatrix} = \begin{pmatrix} C & DF \\ 0 & F \end{pmatrix} \begin{pmatrix} x(t) \\ z(t) \end{pmatrix}.
\end{cases}
\tag{4.12}
$$

Of particular interest are compensators where the state vector $z(t)$ of the controller can be seen as an estimate of the state vector $x(t)$ of the system. Define the estimation error by

$$
e(t) = x(t) - z(t).
$$

The quality of the state estimate can be measured by comparing the predicted output $Cz(t)$ with the observed output $y(t) = Cx(t)$. Now $z(t)$ is supposed to be an estimate of the state. Hence, it is natural to expect it to satisfy a dynamic relation of the form $z(t + 1) = Az(t) + Bu(t) + f(t)$, where $f(t)$ denotes an error term that should be based on the quality of the state estimate, that is, on $y(t) - Cz(t) - Du(t)$. This suggests to choose the following particular form of the state dynamics in the compensator (4.11):

$$
z(t + 1) = Az(t) + Bu(t) + R(y(t) - Cz(t) - Du(t)),
\tag{4.13}
$$

for some matrix R. In this case the error dynamics is given by

$$
e(t + 1) = (A - RC)e(t).
\tag{4.14}
$$

The system (4.13) is called a *state observer* for the system (4.11) if $e(t) \to 0$ for $t \to \infty$, for all initial values $x(0)$ and $z(0)$. So, state observers are characterized by the condition that $A - RC$ is stable.

Definition 4.4.1 The pair (A, C) is called *detectable* if there exists a matrix R such that $A - RC$ is stable.

Theorem 4.4.2 (i) *The pair (A, C) is detectable if and only if every (A, C)-unobservable eigenvalue of A lies in the open unit disc.*
(ii) *There exists a state observer (4.13) for the system (4.6) if and only if the pair (A, C) is detectable.*

Proof   (i) This is the dual version of Theorem 4.3.1. Indeed, the pair (A, C) is detectable if and only if the pair (A^T, C^T) is stabilizable.

(ii) This is evident from the definitions.   □

Next we consider conditions for the existence of a stabilizing compensator.

Theorem 4.4.3   (i) *The system* (4.6) *can be stabilized by a compensator* (4.11) *if and only if* (A, B) *is stabilizable and* (A, C) *is detectable.*
(ii) *Let* R *and* F *be such that* $A - RC$ *and* $A + BF$ *are stable matrices. Then a stabilizing compensator is given by* (4.11) *with state dynamics* (4.13), *that is,* $M = A - RC$, $N = B - RD$ *and* $L = R$.

Proof   First assume that (A, B) is stabilizable and (A, C) is detectable. Let R and F be such that $A - RC$ and $A + BF$ are stable. Then the closed loop system (4.12) obtained by taking (4.11) as compensator with state dynamics (4.13) can be obtained as follows for $t = 0, 1, \ldots$:

$$\Sigma_{cl} \begin{cases} x(t+1) & = Ax(t) + Bu(t), \\ y(t) & = Cx(t) + Du(t), \\ u(t) & = Fz(t), \\ z(t+1) & = Az(t) + Bu(t) + R(y(t) - \bar{y}(t)), \\ \bar{y}(t) & = Cz(t) + Du(t). \end{cases}$$

Here:

$x(t)$ is the unknown state of the system,
$y(t)$ is the measured output of the system,
$z(t)$ is the known estimated state of the system,
$u(t)$ is the computed input of the system,
$\bar{y}(t)$ is the known estimate of the output based on the estimated state.

By eliminating $u(t)$, $y(t)$ and $\bar{y}(t)$ from Σ_{cl} one finds for the state $(x(t)^T \ z(t)^T)^T$ the following for $t = 0, 1, \ldots$:

$$\Sigma_{cl} \begin{cases} x(t+1) & = Ax(t) + BFz(t), \\ z(t+1) & = RCx(t) + (A - RC + BF)z(t). \end{cases}$$

Thus the state transition matrix is

$$\begin{pmatrix} A & BF \\ RC & A - RC + BF \end{pmatrix}.$$

If we compute the state transition matrix in terms of the transformed state variables $(x(t)^T \ e(t)^T)^T = (x(t)^T \ x(t)^T - z(t)^T)^T$ we obtain

$$\begin{pmatrix} I & 0 \\ I & -I \end{pmatrix} \begin{pmatrix} A & BF \\ RC & A - RC + BF \end{pmatrix} \begin{pmatrix} I & 0 \\ I & -I \end{pmatrix} = \begin{pmatrix} A + BF & -BF \\ 0 & A - RC \end{pmatrix}.$$

This is a stable matrix. This proves part (ii) and the "if" part of (i).

It remains to prove that stabilizability of the system implies that the pair (A, B) is stabilizable and the pair (A, C) is detectable. Stabilizability means that $\begin{pmatrix} A & BF \\ LC & M + (N + LD)F \end{pmatrix}$ is stable. Let λ be an (A, C)-unobservable eigenvalue, and let $x \neq 0$ be such that $Ax = \lambda x$, $Cx = 0$. Then

$$\begin{pmatrix} A & BF \\ LC & M + (N + LD)F \end{pmatrix} \begin{pmatrix} x \\ 0 \end{pmatrix} = \lambda \begin{pmatrix} x \\ 0 \end{pmatrix},$$

and as this matrix is stable it follows that $|\lambda| < 1$. According to Theorem 4.4.3 (i) the pair (A, C) is detectable. Now let λ be an (A, B)-uncontrollable eigenvalue and let $x^T \neq 0$ be such that $x^T A = \lambda x^T$, $x^T B = 0$. Then

$$\begin{pmatrix} x^T & 0 \end{pmatrix} \begin{pmatrix} A & BF \\ LC & M + (N + LD)F \end{pmatrix} = \lambda \begin{pmatrix} x^T & 0 \end{pmatrix}.$$

Again we conclude $|\lambda| < 1$ and according to Theorem 4.3.3 the pair (A, B) is stabilizable. $\square$

The result in part (ii) of the above theorem is called the *separation principle*. It shows that the stabilization problem with unobserved state can be solved in two independent steps. First the unobserved state $x(t)$ is estimated by $z(t)$ by a state observer of the form (4.13), by choosing R such that $A - RC$ is stable. Then the controller $u(t) = Fz(t)$ is applied, with F chosen as in the case of observed states, that is, with $A + BF$ stable. This separation of estimation and control is possible for linear input-output systems. Later we will obtain a similar result for linear stochastic input-output systems. It should be mentioned that the separation principle may fail to produce stability for more complex systems.

Example 4.4.1 Continuing our analysis of Example 4.3.1, the system (4.6) has input $g(t)$ and outputs $\left(c(t)\ i(t)\ y(t) \right)^{T}$ with state space matrices

$$A = \begin{pmatrix} \beta(1+\delta) & -\beta\delta \\ 1 & 0 \end{pmatrix}, \ B = \begin{pmatrix} 1 \\ 0 \end{pmatrix}, \ C = \begin{pmatrix} \beta & 0 \\ \beta\delta & -\beta\delta \\ \beta(1+\delta) & -\beta\delta \end{pmatrix}, \ D = \begin{pmatrix} 0 \\ 0 \\ 1 \end{pmatrix}.$$

As $\beta \neq 0$ and $\delta \neq 0$ it follows that rank $C = 2$, so that (A, C) is observable. Also (A, B) is controllable. Therefore the system can be stabilized by output feedback. In Example 4.3.1 we already constructed a matrix F such that $A + BF$ is stable. A matrix R such that $A - RC$ is stable is given, for example, by $R = \begin{pmatrix} 0 & 0 & -1 \\ 0 & 0 & 0 \end{pmatrix}$, as in this case $A - RC = \begin{pmatrix} 0 & 0 \\ 1 & 0 \end{pmatrix}$. With this choice the state estimation error $e(t) = (A - RC)^{t} e_0 = 0$ for $t \geq 2$. So this gives a dead beat observer.

For observable systems a dead beat observer can always be constructed. This follows from the pole placement theorem, by choosing R such that $A - RC$ has all its eigenvalues equal to zero.

Likewise, when (A, B) is controllable, one may construct a dead beat controller F, being the matrix such that $A + BF$ has all its eigenvalues equal to zero. Although this may seem to be a perfect choice for all purposes, a dead beat controller may not always be the most desirable one. We shall see this in the next chapter, when additional design conditions are considered. However, even from a purely numerical point of view one may see that a dead beat controller can have disadvantages. Indeed, if A is imperfectly known then trying to place all eigenvalues of $A + BF$ at zero may result in loss of accuracy in the computed eigenvalues.

Optimal Control

5

In this chapter we consider quantitative control objectives for rather general systems. The inputs are chosen to minimize a function that expresses the costs associated with the system evolution. This can be solved by dynamic programming. We pay special attention to the so-called LQ problem, where the system is linear and the cost function is quadratic. In this case the optimal control is given by state feedback, and the feedback matrix can be computed by solving certain matrix equations (so called Riccati equations).

5.1 Problem Statement

Whereas stability is a qualitative property, in many control applications one is also interested in the quantitative performance of the system. For instance, one could wish to keep the outputs close to a desired trajectory. In most situations the application of control inputs will be associated with costs, for example in terms of energy or money. We assume that these control objectives can be expressed in a single cost function. The resulting optimal control problem is analysed first in a general setting, and then for the case of linear systems with quadratic cost functions. In the latter case the optimal control law can be obtained simply in terms of the parameters of a state space realization of the system. Throughout this chapter we assume that the system is given in state space form and that the state is observed, so that state feedback controllers can be applied.

The general optimal control problem is formulated as follows. The system is described by

$$x(t+1) = f_t(x(t), u(t)), \qquad x(0) = x_0 \text{ given.} \tag{5.1}$$

© The Author(s), under exclusive license to Springer Nature Switzerland AG 2021
C. Heij et al., *Introduction to Mathematical Systems Theory*,
https://doi.org/10.1007/978-3-030-59654-5_5

Here f_t is a function of states and inputs with values in the state space. Note that this function may also depend on t, a fact which is expressed in the notation by the subscript t. We are interested in the behaviour of the system for a fixed set of times $t = 0, 1, \ldots, N$. In addition to the system we have cost functions at each time instant t, which we denote by k_t. The function k_t is a scalar valued function of the state and the input at time t, except for the function k_N which we assume to depend only on the state at time N. The objective is to minimize the total cost function

$$J(x_0, u) = k_N(x(N)) + \sum_{t=0}^{N-1} k_t(x(t), u(t)), \tag{5.2}$$

over all trajectories $u(t) = (u(0), \ldots, u(N-1))$, where the state evolution is controlled by (5.1). So we should choose the inputs $\{u(t) \mid t = 0, \ldots, N-1\}$ such that this cost is minimized, for given initial state x_0. Hence k_N expresses the final cost, and k_t the combined cost of control and system performance at time t. If $N < \infty$ this is called a finite horizon problem. If the term $k_N(x(N))$ is dropped from (5.2), and the summation in the second term runs up to $N = \infty$, then this is an infinite horizon problem. In the latter case the control problem only makes sense if the inputs can be chosen so that the total cost is finite.

As a particular case of special interest we consider the so-called LQ problem, where the system (5.1) is linear and the cost function (5.2) is quadratic. For simplicity we describe only the time invariant case, for which the finite horizon problem is to minimize

$$J(x_0, u) = x(N)^T M x(N) + \sum_{t=0}^{N-1} (x(t)^T Q x(t) + u(t)^T R u(t)) \tag{5.3}$$

subject to

$$x(t+1) = Ax(t) + Bu(t), \quad x(0) = x_0 \text{ given}. \tag{5.4}$$

Here M, Q and R are symmetric matrices with M and Q positive semidefinite and R positive definite. So the objective is to keep both the states and the inputs small. The interpretation in many applications is that the system describes the deviation from a desired trajectory. Further, as R is positive definite this means that every control action gives rise to costs. The LQ problem is of much practical relevance, because it leads to a relatively simple optimal control law.

Example 5.1.1 Consider a trader on a single commodity market. Every day the trader can buy or sell the commodity. We assume that also short selling is possible, that is, the trader can sell more than he owns right now. Further we assume that the net amount bought or sold per day is limited. Let $u(t)$ denote the amount sold on day t, negative values meaning that the trader bought the commodity. The trade restriction is formulated as

$$|u(t)| \leq K, \qquad t = 0, 1, \ldots, N.$$

Let $m(t)$ denote the amount of money and $g(t)$ the amount of goods owned by the trader at day t. By $p(t)$ we denote the price per unit of the good at day t. Further, let s be the cost per day to keep one unit of the good in portfolio. Starting with the initial capital $m(0) + p(0)g(0)$, the goal of the trader is to maximize the final capital

$$m(N) + p(N)g(N).$$

The amount of money and goods evolves according to

$$m(t + 1) = m(t) + p(t)u(t) - sg(t),$$

$$g(t + 1) = g(t) - u(t).$$

In a situation of perfect foresight, the price trajectory $\{p(t) \mid t = 0, \ldots, N\}$ is known. Defining the state variable $x(t) = \begin{pmatrix} m(t) & g(t) \end{pmatrix}^T$, this problem fits in the general formulation (5.1), (5.2). The final cost is $k_N(x(N)) = -m(N) - p(N)g(N)$, and $k(t) = 0$ for $t = 0, \ldots, N - 1$.

Example 5.1.2 We return to Example 1.1.2. Suppose that the government aspires to bring consumption, income and investment close to their equilibrium values expressed in (1.29), for a given equilibrium value $\overline{G}$ of government spending. In state space form the deviations from equilibrium are described by (1.37), (1.38), that is, with state variable $z(t) = \begin{pmatrix} y(t-1) & y(t-2) \end{pmatrix}^T$, input $u(t) = g(t)$ and with state space parameters

$$A = \begin{pmatrix} \beta(1+\delta) & -\beta\delta \\ 1 & 0 \end{pmatrix}, \quad B = \begin{pmatrix} 1 \\ 0 \end{pmatrix}, \quad C = \begin{pmatrix} \beta & 0 \\ \beta\delta & -\beta\delta \\ \beta(1+\delta) & -\beta\delta \end{pmatrix}, \quad D = \begin{pmatrix} 0 \\ 0 \\ 1 \end{pmatrix}.$$

As objective function the government could consider

$$J(z_0, g) = \sum_{t=0}^{N-1} (k_1 c(t)^2 + k_2 i(t)^2 + k_3 y(t)^2 + k_4 g(t)^2)$$

with $k_i > 0$, $i = 1, 2, 3, 4$, the relative costs of deviations from equilibrium for each of the variables. Using (1.38) it follows that this cost function is quadratic and given by

$$J(z_0, g) = \sum_{t=0}^{N-1} (z(t)^T Q z(t) + g(t)^T R g(t) + 2x(t)^T S g(t))$$

with cost parameters given by

$$Q = C^T \operatorname{diag}(k_1, k_2, k_3)C, \quad R = k_3 + k_4, \quad S = C^T \begin{pmatrix} 0 \\ 0 \\ k_3 \end{pmatrix}.$$

So this is a slightly more general LQ problem where (5.3) is extended with the cross term $2z(t)^T Su(t)$. (See also (5.7) below.)

5.2 Dynamic Programming

We consider the dynamical system (5.1) with cost function (5.2). The optimal control problem is to minimize (5.2) by choosing the input sequence $\{u(t) \mid t = 0, \ldots, N-1\}$. We restrict the attention to static state feedback controllers, so that

$$u(t) = h_t(x(t)),$$

for some function h_t. The input set may also depend on the current state of the system, so that $u(t) \in U_t(x(t))$.

Dynamic programming gives a recursive solution for this optimal control problem. The idea is as follows. Let $\{u^*(t) \mid t = 0, \ldots, N-1\}$ be the optimal solution, and let $\{x^*(t) \mid t = 0, \ldots, N-1\}$ be the resulting state trajectory generated by (5.1). At time $t = i$, the inputs $\{u^*(t) \mid t = i, \ldots, N-1\}$ will then minimize the costs

$$k_N(x(N)) + \sum_{t=i}^{N-1} k_t(x(t), u(t)), \qquad x(i) = x^*(i).$$

Indeed, if another input $\{u^+(t) \mid t = i, \ldots, N-1\}$ would give smaller costs, then the input defined by $u(t) = u^*(t)$ for $t \leq i - 1$ and $u(t) = u^+(t)$ for $t \geq i$ would give a cost (5.2) that is smaller than that obtained by taking the inputs $\{u^*(t) \mid t = 0, \ldots, N-1\}$. This contradicts optimality of the inputs $\{u^*(t) \mid t = 0, \ldots, N-1\}$. This is called the *principle of optimality*, or *Bellman's principle of optimality*. It leads to the following dynamic programming solution, in terms of the *cost-to-go functions* $J_t : \mathbb{R}^n \to \mathbb{R}$ defined recursively by

$$J_N(x) = k_N(x), \tag{5.5}$$

$$J_t(x) = \min_{u \in U_t(x)} (k_t(x, u) + J_{t+1}(f_t(x, u))). \tag{5.6}$$

Theorem 5.2.1 *For given* $x \in \mathbb{R}^n$ *let* $u = h_t(x)$ *be a minimizing input for* (5.6). *Then the control input defined by*

$$u^*(t) = h_t(x^*(t)), \quad x^*(t + 1) = f_t(x^*(t), u^*(t)), \quad x^*(0) = x_0$$

minimizes (5.2) *and the minimal cost is equal to* $J(x_0, u^*) = J_0(x_0)$.

Proof The minimization of (5.2) can be written as

$$\min_{u(0),\dots,u(N-1)} J(x_0, u) = \min_{u(0),\dots,u(N-1)} \{k_0(x_0, u(0)) + k_1(x(1), u(1)) +$$

$$+ \cdots + k_{N-1}(x(N - 1), u(N-1)) + k_N(x(N))\}.$$

As only the last two terms depend on $u(N - 1)$, we may write this as

$$\min_{u(0),\dots,u(N-2)} \Big[\{k_0(x_0, u(0)) + \cdots + k_{N-2}(x(N - 2), u(N - 2))\} +$$

$$+ \min_{u(N-1)} \{k_{N-1}(x(N - 1), u(N - 1)) + k_N(x(N))\}\Big].$$

Using $k_N(x(N)) = J_N(x(N)) = J_N(f_{N-1}(x(N - 1), u(N - 1)))$ it follows that

$$\min_{u(N-1)} \{k_{N-1}(x(N - 1), u(N - 1)) + k_N(x(N))\}$$

$$= \min_{u(N-1)} \{k_{N-1}(x(N - 1), u(N - 1)) + J_N(f_{N-1}(x(N - 1), u(N - 1)))\}$$

$$= J_{N-1}(x(N - 1)),$$

so that $u^*(N - 1) = h_{N-1}(x(N - 1))$ is optimal and

$$\min_{u(0),\dots,u(N-1)} J(x_0, u)$$

$$= \min_{u(0),\dots,u(N-2)} \Big[k_0(x_0, u(0)) + \cdots$$

$$+ k_{N-2}(x(N - 2), u(N - 2)) + J_{N-1}(x(N - 1))\Big].$$

Again, as only the last two terms depend on $u(N-2)$, this can be written as

$$\min_{u(0),\dots,u(N-1)} J(x_0, u)$$

$$= \min_{u(0),\dots,u(N-3)} \Big[\{k_0(x_0, u(0)) + \dots + k_{N-3}(x(N-3), u(N-3))\} +$$

$$+ \min_{u(N-2)} \{k_{N-2}(x(N-2), u(N-2)) + J_{N-1}(x(N-1))\} \Big],$$

where

$$\min_{u(N-2)} \{k_{N-2}(x(N-2), u(N-2)) + J_{N-1}(x(N-1))\}$$

$$= \min_{u(N-2)} \{k_{N-2}(x(N-2), u(N-2)) + J_{N-1}(f_{N-2}(x(N-2), u(N-2)))\}$$

$$= J_{N-2}(x(N-2)).$$

This shows that $u^*(N-2) = h_{N-2}(x(N-2))$ is optimal, and continuing in this fashion shows that the algorithm of the theorem indeed gives minimum costs. $\square$

This algorithm can only be applied for finite horizons, as the recursions (5.5), (5.6) start at the final time $t = N$. The infinite horizon problem requires techniques that are beyond the scope of this book. See for instance [7] for an excellent book on dynamic programming.

Although in the other parts of the book we restrict attention to linear systems we have chosen to treat here the general problem of minimizing (5.2) subject to (5.1). The reason for this is that the argument in the proof of Theorem 5.2.1 works just as well and probably even more transparent in the more general case discussed here.

Example 5.2.1 Consider the trader described in Example 5.1.1. It is assumed that the price trajectory $\{p(t) \mid t = 0, \dots, N\}$ is known to the trader. Before giving a formal analysis, common sense suggests the following strategy. On day t, the trader has to decide whether to buy or to sell the good. This decision has no effect on his future options, as the only restriction is that $|u(t)| \le K$ on all days. Therefore he should buy if the final profit $p(N)$ is larger than the cost $p(t) + (N-t-1)s$ associated with buying the good now and keeping it until time N. He should sell if $p(N) < p(t) + (N-t-1)s$.

To prove the optimality of this strategy we solve the optimization problem explicitly. In this simple case this can be done directly by writing out the objective function

$J((m_0, g_0), u) = -m(N) - p(N)g(N)$. Indeed, there holds $g(N) = g_0 - \sum_{t=0}^{N-1} u(t)$
and

$$m(N) = m_0 + \sum_{t=0}^{N-1} (p(t)u(t) - sg(t))$$

$$= m_0 + \sum_{t=0}^{N-1} p(t)u(t) - s \sum_{t=0}^{N-1} \left(g_0 - \sum_{i=0}^{t-1} u(i)\right)$$

$$= m_0 - Nsg_0 + \sum_{t=0}^{N-1} p(t)u(t) + s \sum_{t=0}^{N-1} (N - t - 1)u(t),$$

so that the cost function is given by

$$J((m_0, g_0), u) = -(m_0 + (p(N) - sN)g_0) + \sum_{t=0}^{N-1} (p(N) - p(t) - (N - t - 1)s)u(t).$$

This is minimized by choosing $u(t) = K$, that is, to sell the maximal amount, if $p(t) + (N-t-1)s > p(N)$, and to buy the maximal amount, $u(t) = -K$, if $p(t)+(N-t-1)s < p(N)$.

The optimality can be checked also by the dynamic programming algorithm. The cost functions are given by $k_N(m, g) = -m - p(N)g$ and $k_t = 0$ for $t = 0, \ldots, N - 1$. The first step is to solve (5.6) for $t = N - 1$, by chosing $u(N - 1)$ equal to the value u in $[-K, K]$ that minimizes

$$J_N(m(N - 1) + p(N - 1)u(N - 1) - sg(N - 1), g(N - 1) - u(N - 1))$$

$$= -(m(N - 1) + p(N - 1)u(N - 1) - sg(N - 1)) -$$

$$- p(N)(g(N - 1) - u(N - 1))$$

$$= -(m(N - 1) + (p(N) - s)g(N - 1)) + u(N - 1)(p(N) - p(N - 1)).$$

This shows that $u(N - 1) = K$ if $p(N) \leq p(N - 1)$ and $u(N - 1) = -K$ if $p(N) > p(N - 1)$, with resulting cost-to-go function

$$J_{N-1}(m(N - 1), g(N - 1))$$

$$= -(m(N - 1) + (p(N) - s)g(N - 1) + K|p(N) - p(N - 1)|).$$

The next step is to choose $u(N-2)$ equal to the $u \in [-K, K]$ that minimizes

$$
\begin{aligned}
J_{n-2}\big(m(N-1), g(N-1)\big) &= -\big(m(N-2) - sg(N-2) + p(N-2)u + \\
&\quad + \big(p(N) - s\big)\big(g(N-2) - u\big) + K|p(N) - p(N-1)| \\
&= \big(m(N-2) - (p(N) - 2s)g(N-2)\big) - \big(p(N) - p(N-2) - s\big)u.
\end{aligned}
$$

Hence $u_{N-2} = K$ if $p(N) - p(N-2) - s < 0$ and $u_{N-2} = -K$ otherwise. Continuing in this way shows the optimality of the above strategy, with cost-to-go functions

$$
J_t(m(t), g(t))
$$

$$
= -(m(t) + [p(N) - (N-t)s]g(t) + K \sum_{i=t}^{N-1} |p(N) - p(i) - (N-i-1)s|).
$$

Note that the optimal strategy leads to extreme actions, that is, always the maximally allowed amount is bought or sold. This is a widespread phenomenon, called *bang-bang control*. Further, the strategy requires perfect foresight. In practice the price trajectory, and in particular the final price $p(N)$, will not be known when the trader has to make his decisions on days $t < N$. If the price is uncertain then this decision problem also becomes uncertain. Later we describe stochastic systems to model this kind of uncertainty. The optimal strategy can then be determined by stochastic dynamic programming.

5.3   Linear Quadratic Control

Dynamic programming is a very general technique. The principal condition for practical applications is that the minimization in (5.6) should be solved as a function of the state variable x, that is, we need to determine the feedback functions $h_t(x)$ in Theorem 5.2.1. In this section we show that for the LQ problem these feedbacks can be computed in a simple way.

Let the system be linear as in (5.4), that is

$$
x(t+1) = Ax(t) + Bu(t), \quad x(0) = x_0 \text{ given.}
$$

We consider a generalization of the cost function (5.3), that is,

$$
\begin{aligned}
J(x_0, u) &= x^T(N)Mx(N) + \\
&\quad + \sum_{t=0}^{N-1} x^T(t)Qx(t) + u^T(t)Ru(t) + u^T(t)Sx(t) + x^T(t)S^Tu(t).
\end{aligned}
$$

(5.7)

We impose the conditions that both matrices $\begin{pmatrix} Q & S^T \\ S & R \end{pmatrix}$ and M are positive semidefinite and that R is positive definite. Note that $u^T(t)Sx(t) + x^T(t)S^T u(t) = 2u^T(t)Sx(t)$, so (5.7) can be rewritten as

$$J(x_0, u) = x^T(N)Mx(N) + \sum_{t=0}^{N-1} x^T(t)Qx(t) + u^T(t)Ru(t) + 2u^T(t)Sx(t).$$

The LQ optimal control problem is to minimize (5.7) for a given system (5.4).

Before solving this optimal control problem, we comment on another form of cost function that is useful, and that will appear later. Consider a cost function of the form

$$J_1(x_0, u) = x^T(N)Mx(N) + \sum_{t=0}^{N-1} \|R_1 x(t) + S_1 u(t)\|^2, \tag{5.8}$$

where we impose that conditions that M is positive semidefinite and that $S_1^T S_1$ is positive definite. Obviously, this type of cost function can be viewed as a special case of (5.7), with $R = S_1^T S_1$, $Q = R_1^T R_1$, and $S = S_1^T R_1$. Conversely, every cost function (5.7) can also be rewritten in the form (5.8), by taking $R_1 = \begin{pmatrix} Q^{1/2} \\ R^{-1/2}S \end{pmatrix}$ and $S_1 = \begin{pmatrix} 0 \\ R^{1/2} \end{pmatrix}$. Thus, results we shall present in the sequel for the cost function of the form (5.7) can be reformulated in terms of the equivalent cost function (5.8).

The solution of the LQ optimal control problem is given by dynamic programming, and has the following form.

Theorem 5.3.1 *The optimal control law for the LQ problem (5.4), (5.7) is given by the state feedback law*

$$u^*(t) = h_t(x(t)) = F_t x(t) \tag{5.9}$$

where

$$F_t = -(R + B^T K_{t+1} B)^{-1}(S + B^T K_{t+1} A). \tag{5.10}$$

The minimal cost is given by

$$J(x_0, u^*) = x_0^T K_0 x_0, \tag{5.11}$$

*where the matrices K_t are defined by the backwards recursion given by the Riccati
difference equation*

$$
\begin{cases}
K_t & = Q + A^T K_{t+1} A - F_t^T (R + B^T K_{t+1} B) F_t, \\
F_t & = -(R + B^T K_{t+1} B)^{-1} (S + B^T K_{t+1} A), \\
K_N & = M.
\end{cases}
\tag{5.12}
$$

Moreover, the cost-to-go functions J_t are given by $J_t(x) = x(t)^T K_t x(t)$.

Proof We apply Theorem 5.2.1, with the cost functions $k_N(x) = x^T M x$ and $k_t(x, u) = x^T Q x + u^T R u + 2u^T S x$. First, we will prove that $J_t(x) = x^T K_t x$ with J_t as defined in
(5.6) and K_t as in (5.12). This is evidently correct for $t = N$.

We proceed by induction and suppose that $J_{t+1}(x) = x^T K_{t+1} x$. First remark that
cost functions $k_t \geq 0$ for $t = 0, \dots, N$. Hence the cost to go function $J_{t+1}(x)$ also has
$J_{t+1}(x) \geq 0$ for all $x \in \mathbb{R}^n$. So K_{t+1} is positive semidefinite, and R is positive. This
implies that $R + B^T K_{t+1} B$ is invertible. We see that F_t is well defined by (5.12). Now we
prove that $J_t(x) = x^T K_t x$, with the minimizing input given by (5.9). In (5.6) the following
expression should be minimized:

$$
k_t(x, u) + J_{t+1}(f_t(x, u)) = x^T Q x + u^T R u + 2u^T S x + J_{t+1}(Ax + Bu)
$$

$$
= x^T Q x + u^T R u + 2u^T S x + (Ax + Bu)^T K_{t+1}(Ax + Bu)
$$

$$
= x^T (Q + A^T K_{t+1} A) x +
$$

$$
+ u^T (R + B^T K_{t+1} B) u - 2u^T (R + B^T K_{t+1} B) F_t x.
$$

By completing the squares the above expression can be rewritten as

$$
(u - F_t x)^T (R + B^T K_{t+1} B)(u - F_t x) +
$$

$$
+ x^T (Q + A^T K_{t+1} A) x -
$$

$$
- (F_t x)^T (R + B^T K_{t+1} B)(F_t x).
$$

As $R + B^T K_{t+1} B$ is positive definite, this expression is minimized by taking the input (5.9)
with resulting cost-to-go $J_t(x) = x^T K_t x$, where K_t is given by (5.12). This concludes the
inductive proof.

Finally, according to Theorem 5.2.1 the minimal cost is given by $J_0(x_0) = x_0^T K_0 x_0$. $\square$

From the theorem one sees that the LQ controller is given by linear state feedback.
This makes LQ control attractive, as this controller can be implemented relatively easily.

The feedback gains in (5.9) depend on time, but they can be computed recursively and independent of the actual observations of the system. One says that the control problem is solved off-line. That is, we only need to know the system (5.4) and the cost parameters in (5.7), to compute the optimal feedback matrices in (5.9), by the recursions (5.12). So, the controller can be constructed before the actual observations are available.

The result of Theorem 5.3.1 is true as well for time-varying linear systems with time varying cost function. That is, a similar result holds when the objective is to minimize the cost function $J(x_0, u) = \sum_{t=0}^{N-1}(x(t)^T Q(t)x(t) + u(t)^T R(t)u(t) + 2x(t)^T S(t)u(t)) + x(N)^T M x(N)$, where $Q(t) = Q(t)^T$ and $R(t) = R(t)^T > 0$, subject to $x(t + 1) = A(t)x(t) + B(t)u(t)$.

In some applications it may not be so clear how to choose the horizon N. This motivates study of the *infinite horizon problem* with cost function

$$J_\infty(x_0, u) = \sum_{t=0}^{\infty} x^T(t)Qx(t) + u^T(t)Ru(t) + 2u^T(t)Sx(t), \tag{5.13}$$

or the infinite horizon analogue of (5.8) given by

$$I_{1,\infty}(x_0, u) = \sum_{t=0}^{\infty} \| R_1 x(t) + S_1 u(t) \|^2, \tag{5.14}$$

subject to $x(t + 1) = Ax(t) + Bu(t)$ with $x(0) = x_0$ given. Of course, this only makes sense if the cost can be given a finite value by appropriate choice of the inputs. A sufficient condition for this is that the pair (A, B) is stabilizable.

One could expect that the infinite horizon problem can be approximated by considering finite horizon problems with horizon $N \to \infty$. Under suitable conditions this idea indeed works. Moreover, the solutions of the Riccati difference equations (5.12) then converge to a limit, which is independent of the choice of the final cost M, and the corresponding control law (5.9) becomes time invariant in the limit. So the solution of the infinite horizon problem becomes quite simple in this case.

Theorem 5.3.2 *Assume that the pair $(A - BR^{-1}S, Q - S^T R^{-1}S)$ is detectable and that the pair (A, B) is stabilizable. Denote the solutions of (5.12) at $t = 0$, obtained by starting at $K_N = M$, with M positive definite, by $K_0(N, M)$. Then the following hold true:*

(i) $\lim_{N\to\infty} K_0(N, M) = K_+$ *exists and is independent of M. Denote*

$$F_+ = -(R + B^T K_+ B)^{-1}(B^T K_+ A + S). \tag{5.15}$$

(ii) K_+ *satisfies the algebraic Riccati equation*

$$K = Q + A^T K A - (A^T K B + S^T)(R + B^T K B)^{-1}(B^T K A + S). \qquad (5.16)$$

Moreover, it is the largest Hermitian solution of (5.16), *in the sense that for any other Hermitian solution K the matrix $K_+ - K$ is positive semidefinite. Furthermore, it is also the unique solution K for which $A - B(R + B^T K B)^{-1}(B^T K A + S)$ is a stable matrix. In particular, $A + B F_+$ is a stable matrix. If, in addition, the pair $(A - B R^{-1} S, Q - S^T R^{-1} S)$ is observable, then K_+ is positive definite.*

(iii) *The optimal control law for* (5.4), (5.13) *is given by*

$$u^*(t) = h_t(x) = F_+ x(t), \qquad (5.17)$$

with cost $J_\infty(x_0, u^*) = x_0^T K_+ x_0$.

Proof   We shall not prove all statements of the theorem, but we will concentrate our attention on the ones that are most relevant for our purpose. In particular, we will prove only part of the statements in (i) and (ii). See [1] for full proofs.

The reader should be forewarned that the pace of the arguments in this proof is considerably higher than the pace usual for the preceding parts of the book.

As (A, B) is stabilizable there exists a matrix F such that $A + B F$ is stable, that is, all eigenvalues of $A + B F$ are inside a circle of radius less than one. Hence if we employ F as feedback and take $u(t) = F x(t)$, then the state $x(t)$ is given by $x(t) = (A + B F)^t x_0$. Note that $\|(A + B F)^t\| \le C_0 r^t$ for some constant C_0 and some $0 < r < 1$. Hence $\|x(t)\|$ can be estimated as follows: $\|x(t)\| \le C \cdot r^t$ for some constant C. But then also

$$\left\| \begin{pmatrix} x(t) \\ u(t) \end{pmatrix} \right\| = \left\| \begin{pmatrix} x(t) \\ F x(t) \end{pmatrix} \right\| = \left\| \begin{pmatrix} I \\ F \end{pmatrix} (A + B F)^t x_0 \right\| \le C_1 \cdot r^t,$$

for some constant C_1. Hence for this choice of inputs the cost function $J_\infty(x_0, u)$ is finite for each choice of x_0. This implies that the optimal cost, $\inf_u J_\infty(x_0, u)$ is finite.

Let $M = 0$ in (5.12) and denote k_t for $t = 0$ by $K_0(N)$. Let us denote by u^* the input minimizing $J_\infty(x_0, u)$ and by u_n^* the input minimizing $J_n(x_0, u)$. Then

$$\inf_u J_N(x_0, u) \le \sum_{t=0}^{N-1} \left(u^*(t)^T \ \ x^*(t)^T \right) \begin{pmatrix} Q & S^T \\ S & R \end{pmatrix} \begin{pmatrix} u^*(t) \\ x^*(t) \end{pmatrix} \le$$

$$\le \sum_{t=0}^{\infty} \left(u^*(t)^T \ \ x^*(t)^T \right) \begin{pmatrix} Q & S^T \\ S & R \end{pmatrix} \begin{pmatrix} u^*(t) \\ x^*(t) \end{pmatrix} = \inf_u J_\infty(x_0, u)$$

This implies that $x_0^* K_0(N) x_0 \leq \inf_u J_\infty(x_0, u)$. Therefore, the sequence $\{K_0(N)\}_{N=1}^\infty$ has an upper bound.

The same reasoning applies to show that $K_0(N) \leq K_0(N+1)$. Thus the sequence $\{K_0(N)\}_{N=0}^\infty$ is a increasing sequence of positive semidefinite matrices that is bounded above. Thus there is a limit, which we denote by K_+ for the time being. As $K_0(N)$ satisfies the Riccati difference equation (5.12) (with $M = 0$) it is immediate that K_+ satisfies the algebraic Riccati equation (5.16). Moreover, it is clear that $0 \leq x_0^T K_+ x_0 \leq \inf_u J_\infty(x_0, u)$.

To show that the corresponding closed loop feedback matrix $A + BF_+$ is stable, we can rewrite (5.16) for $K = K_+$ as follows, where we use for convenience the notation $A_+ = A + BF_+$:

$$K_+ = Q + A^T K_+ A + (A^T K_+ B + S^T) F_+ =$$
$$= Q + A_+^T K_+ A_+ - F_+^T (B^T K_+ A + B^T K_+ B F_+) + S^T F_+ =$$
$$= Q + A_+^T K_+ A_+ + F_+^T (R F_+ + S) + S^T F_+ =$$
$$= A_+^T K_+ A_+ + \begin{pmatrix} I & F_+^T \end{pmatrix} \begin{pmatrix} Q & S^T \\ S & R \end{pmatrix} \begin{pmatrix} I \\ F_+ \end{pmatrix}.$$

This shows that $K_+ - A_+^T K_+ A_+$ is positive semidefinite. Since (A, B) is stabilizable and $(A - BR^{-1}S, Q - S^T R^{-1}S)$ is detectable A_+ is stable. Compare Theorem 4.1.4. We shall not provide the details here.

Let us consider the input obtained by taking the static state feedback $u(t) = -(R + B^T K_+ B)^{-1}(S + B^T K_+ A)x(t) = -F_+ x(t)$. The cost associated with this input is equal to $x_0^T K_+ x_0$, as one sees by the same completion of the squares argument as used in the proof of Theorem 5.3.1. Indeed, consider for this choice of input

$$J_\infty(x_0, u) + \sum_{t=0}^\infty (x(t+1)^T K_+ x(t+1) - x(t)^T K_+ x(t))$$

$$= -x(0)^T K_+ x(0) + J_\infty(x_0, u)$$

(as A_+ is a stable matrix). On the other hand, using $x(t+1) = Ax(t) + Bu(t)$, and the definition of $J_\infty(x_0, u)$, this is equal to the following, where we suppress the dependence on t for the sake of clarity:

$$\sum_{t=0}^\infty (x^T A^T K_+ A x - x^T K_+ x + x^T Q x$$

$$+ x^T (A^T K_+ B + S^T) u + u^T (S + B^T K_+ A) x + u^T R u + u^T B^T K_+ B u)$$

$$= \sum_{t=0}^{\infty} \{(u + F_+ x)^T (R + B^T K_+ B)(u + F_+ x) +$$

$$+ x^T (-K_+ + A^T K_+ A + Q)x -$$

$$- x^T (A^T K_+ B + S^T)(R + B^T K_+ B)^{-1}(B^T K_+ A + S))x\}.$$

As K_+ satisfies (5.16) it follows that

$$J_{\infty}(x_0, u) = x(0)^T K_+ x(0) + \sum_{t=0}^{\infty} \|(R + B^T K_+ B)^{1/2}(u + F_+ x)\|.$$

But then, for the particular choice of inputs $u = -F_+ x$, we have $J_{\infty}(x_0, u) = x_0^T K_+ x_0$. So $\inf_u J_{\infty}(x_0, u) \leq x_0^T K_+ x_0$ and thus $\inf_u J_{\infty}(x_0, u) = x_0^T K_+ x_0$. It follows that the optimal input is given by $u(t) = F_+ x(t)$.   □

The assumptions in Theorem 5.3.2 that the pair (A, B) is stabilizable and that the pair $(A - BR^{-1}S, Q - S^T R^{-1}S)$ is detectable are made to guarantee that the cost function (5.13) makes sense (that is, (5.13) is finite for some control inputs u), and that the optimal control law (5.17) actually stabilizes the system. Without these assumptions there could exist inputs that minimize (5.13), but which are not stabilizing. Such inputs do not have much practical significance.

As an example, the assumptions of Theorem 5.3.2 are satisfied for cost functions of the form

$$J(x_0, u) = \sum_{t=0}^{\infty} (y(t)^T Q_1 y(t) + u^T(t) R u(t))$$

with Q_1 and R positive definite and with $y(t) = Cx(t)$ the output of the system (5.4) with (A, B, C) a minimal realization. Indeed, in that case (A, B) is controllable and hence stabilizable. As concerns the detectability, clearly in (5.13) we have $S = 0$ and $Q = C^T Q_1 C$, so detectability of (A, Q) follows from observability of (A, C) and positivity of Q_1.

Let us formulate also the main results of Theorem 5.3.2 for the alternative infinite horizon cost function (5.14). Assume that the pair (A, B) is stabilizable, that $S_1^T S_1$ is positive definite, and that the pair $(A - B(S_1^T S_1)^{-1} S_1^T R_1, R_1 - S_1(S_1^T S_1)^{-1} S_1^T R_1)$ is detectable. Consider the algebraic Riccati equation

$$K = R_1^T R_1 + A^T K A - (S_1^T R_1 + B^T K A)^T (S_1^T S_1 + B^T K B)^{-1}(S_1^T R_1 + B^T K A)$$

together with the control law $u(t) = Fx(t)$, where the feedback matrix F is given by $F = -(S_1^T S_1 + B^T K B)^{-1}(S_1^T R_1 + B^T K A)$. If we take for K the unique positive semidefinite solution K_+ of this Riccati equation, then this control law is the optimal one, it is a stabilizing control, and the optimal costs are given by $x_0^T K_+ x_0$.

Example 5.3.1 In Example 5.1.2 we described a control problem for government policy. As an example, suppose that the government expects to be in power for this and the coming 3 years and that the objective is to keep the national income as close as possible to a desired level $\bar{Y}$. Associated with this are equilibrium values of consumption, investment and government expenditures as described in (1.29). Taking all variables in deviation from equilibrium, this gives the linear system described in Example 5.1.2. The new objective function is a special case of the one given in Example 5.1.2 and is given by

$$J(x_0, g) = \sum_{t=0}^{3} y(t)^2,$$

where $x_0 = \left(y(-1)\ y(-2) \right)^T$ and $g = \left(g(0)\ g(1)\ g(2)\ g(3) \right)$. Introduce the matrix $S = \left(\beta(1+\delta)\ -\beta\delta \right)$, then in terms of (5.7), we have $M = 0$, $R = 1$ and $Q = S^T S$. The optimal control follows from Theorem 5.3.1. Here $K_4 = M = 0$ and it follows from (5.12) that also $K_3 = K_2 = K_1 = K_0 = 0$. Therefore, the optimal controller (5.9) is given by

$$g^*(t) = -Rx(t) = -\beta(1+\delta)y(t-1) + \beta\delta y(t-2).$$

Of course, this result could also have been obtained directly from equation (1.33). The minimal cost if given by $x_0^T K_0 x_0 = 0$.

A more realistic cost function would also penalize deviations from equilibrium of the other economic variables, see Example 5.1.2. For given values of β and δ and of the relative weights g_i, $i = 1, 2, 3, 4$, the optimal control law and associated cost are easily obtained by applying Theorem 5.3.1. In practice β and δ are unknown and have to be estimated from observed data. Furthermore, it may be unclear how the relative weights in the cost function should be chosen. Here a sensitivity analysis may be of interest, where the effect of the policy objectives on the economic developments is considered.

For the infinite horizon problem with $J_\infty(x_0, g) = \sum_{t=0}^{\infty} y(t)^2$ the optimal control law is of course the same as before. This can also be checked by means of Theorem 5.3.2. More realistic cost functions require the solution of (5.16).

There exist several reliable algorithms for solving Riccati equations. A simple method is to start with $K_0 = 0$, and to compute K_t, $t \leq -1$ backwards in time using the Riccati difference equation (5.12). For $t \to -\infty$ the matrices K_t converges to the required solution of (5.16). This is never used in practice. A faster method uses an approach via eigenvalues and eigenvectors of certain matrices, and another way to approximate the solution of (5.16) is to apply Newton's method to the nonlinear matrix equation (5.16). However, it would lead us to far afield to explain these methods here in detail. A good source for results on the algebraic Riccati equation is [45].

Stochastic Systems

6

In stochastic systems, the outputs are (partly) driven by unobserved random inputs. This chapter is concerned with stationary processes and their approximation with finite dimensional linear stochastic systems. Similar to the results for deterministic input-output systems there is an equivalence between finite dimensional stochastic state space models, polynomial (ARMA) representations, and rational spectra (in the frequency domain), which are the analogue of the transfer function.

6.1 Modelling

The methods introduced in the foregoing chapters concern the representation and control of completely specified systems. In most applications, however, the precise form of the system is not known. This is the case, for example, in complex technical systems like an airplane or a chemical plant. The situation is even more complicated in economic applications. The reaction mechanisms between economic variables are often known only in qualitative terms, and one cannot identify all factors that influence the system behaviour. This motivates the study of imperfectly known dynamical systems. The uncertainties involved may be modelled in several ways, one of which is the use of stochastic models. System identification is concerned with the construction of a dynamical system for an observed process. The main purpose is to determine the so-called systematic part, which explains the process up to unpredictable variations. In practice one tries to capture these systematic relations by equations of the form

$$G(w, a) = 0. \tag{6.1}$$

Here w denotes trajectories of the observed variables and a denotes auxiliary variables used to facilitate the system description. The auxiliary variables represent unobserved influences. For practical purposes one is interested in relatively simple representations, for example of the form

$$G(t, w(t), w(t-1), \ldots, w(t-L), a(t), a(t-1), \ldots, a(t-L)) = 0. \qquad (6.2)$$

Time invariant models are of particular interest, which means that the function G does not explicitly depend on time. The linearized form of the model consists of linear difference equations of the type

$$F_0 w(t) + \ldots + F_L w(t-L) = B_0 a(t) + \ldots + B_L a(t-L). \qquad (6.3)$$

The essential difference with the models in the foregoing chapters (see for example (1.22)) is that the auxiliary variables $a(t)$ are not observed and that the system parameters $F_0, \ldots, F_L$ and $B_0, \ldots, B_L$ are unknown. If we assume that $F_0 = B_0 = I$ then the term $a(t)$ can be interpreted as the modelling error at time t. For a good specification this model error should not have any predictable dynamical pattern, as this would indicate misspecification. Such a completely unpredictable process is called white noise. This forms the building block for the processes described in this chapter.

Specializing (6.3) to the case of input-output systems with $w = \left(u^T \ y^T \right)^T$, writing $F_i^T = (-C_i^T \ -A_i^T)$, and taking $-A_0 = B_0 = I$ we get the so-called ARMAX *representation*

$$y(t) = A_1 y(t-1) + \ldots + A_L y(t-L) + a(t) + B_1 a(t-1) + \ldots$$
$$+ B_L a(t-L) + C_0 u(t) + C_1 u(t-1) + \ldots + C_L u(t-L) \qquad (6.4)$$

In this model the current value of the output y is explained by an autoregression (AR) on its own past, a moving average (MA) of the auxiliary variables a, and the exogeneous (X) inputs u. Note that this model imposes no restrictions on the input-output behaviour if the variables a are considered as completely arbitrary. For the reason mentioned before it is often supposed that a is a white noise stochastic process. In this chapter we first consider systems without control inputs u, so that the observed process y is driven by white noise. This means that the output y is a stochastic process.

6.2   Stationary Processes

A stochastic process is defined as a collection of random variables $\{y(t), t \in \mathbb{Z}\}$. A time series is an outcome of this process, that is, a series of observed vectors $y(t)$. In practice, the available information often consists of an observed time series and the question is how to estimate the properties of the underlying process. The statistical properties of primary importance are the mean $\mu(t) := E\{y(t)\}$ and the covariances

$$R(t, s) := E\{(y(t) - \mu(t))(y(s) - \mu(s))^T\}.$$

Definition 6.2.1 A stochastic process is (weakly) *stationary* if $\mu(t) = \mu(s)$ and $R(t, s) = R(t + k, s + k)$ for all $t, s, k \in \mathbb{Z}$, that is, if its mean and covariances exist and are time-invariant. In this case we write $\mu(t) = \mu$ and $R(t, s) = R(t - s)$.

In the case of a Gaussian process, that is, a process $\{y(t), t \in \mathbb{Z}\}$ such that the distribution function of each $y(t)$ is multivariable normal, this is equivalent to *strong stationarity* in the sense that the probability distributions are time invariant, that is, for all $n \in \mathbb{N}, k \in \mathbb{Z}, t_i \in \mathbb{Z}, i = 1, \ldots, n$, there holds

$$p(y(t_1), y(t_2), \ldots, y(t_n)) = p(y(t_1 + k), y(t_2 + k), \ldots, y(t_n + k)),$$

where p denotes the joint probability distribution. In the sequel, whenever we say stationary, we mean strongly stationary. Note that for a stationary process we have $R(-k) = R(k)^T$.

In practice the mean and covariances of a stationary process are often not known. If the process is observed on a time interval of length N so that the available data are $y(1), y(2), \cdots, y(N)$, then the *sample mean* is defined by

$$\hat{\mu}_N := \frac{1}{N} \sum_{t=1}^{N} y(t) \tag{6.5}$$

and the *sample covariances* are defined by

$$\hat{R}_N(k) := \frac{1}{N} \sum_{t=k+1}^{N} (y(t) - \hat{\mu}_N)(y(t - k) - \hat{\mu}_N)^T, \quad 0 \le k \le N - 1, \tag{6.6}$$

while for $k < 0$ we have $\hat{R}_N(k) = \hat{R}_N(-k)^T$.

Definition 6.2.2 A stationary stochastic process is *ergodic* if the following holds true almost surely:

$$\lim_{N \to \infty} \hat{\mu}_N = \mu \tag{6.7}$$

$$\lim_{N \to \infty} \hat{R}_N(k) = R(k) \tag{6.8}$$

In the sequel we will simply assume that ergodicity holds true. The sample mean and covariances then provide reliable information on the underlying process if the number of observations is sufficiently large.

Example 6.2.1 A *white noise process* is a stationary process characterized by the property that values at different time instants are uncorrelated, in the sense that $R(k) = 0$ for all $k \neq 0$. The law of large numbers implies that (6.7) holds true. Condition (6.8) requires some additional assumptions, for example, that the process has bounded fourth moments. In case $R(0) = I$ and $\mu = 0$ the white noise process is called *standard white noise*.

A useful evaluation of estimated stochastic systems is to test whether the model errors can be considered as white noise. If this is not the case then additional (linear) dynamical relationships are present in the data.

Example 6.2.2 Consider the univariate process described by the equation

$$y(t) = \sin(\omega t + \theta) \tag{6.9}$$

Suppose that ω is unknown but fixed, and that the phase θ is a random variable with uniform distribution on the interval $[0, 2\pi)$. Then y is a stochastic process. This process is very particular, as it is *perfectly predictable*. Indeed, from the equality $\sin a + \sin b = 2 \sin(\frac{a+b}{2}) \cos(\frac{a-b}{2})$ one can derive

$$y(t + 2) + y(t) = (2 \cos \omega) y(t + 1). \tag{6.10}$$

For every time series generated by the process the value of $\cos \omega$ can be determined from three subsequent observations, and this makes the whole future of the trajectory perfectly predictable. The process y is nonetheless stationary, as it has mean $\mu(t) = 0$ and covariances

$$R(t, s) = \frac{1}{2\pi} \int_0^{2\pi} \sin(\omega t + \theta) \sin(\omega s + \theta) d\theta = \frac{1}{2} \cos(\omega(t - s)) = R(t + k, s + k).$$

This shows that the interdependence between the observations never dies out, even if the distance in time is arbitrarily large.

Example 6.2.3 A *moving average (MA) process* is modelled in terms of uncorrelated driving forces, that is,

$$y(t) = \sum_{k=0}^{\infty} G(k)\varepsilon(t - k) \tag{6.11}$$

where ε is an unobserved standard white noise process. If $\sum_{k=0}^{\infty} \|G(k)\| < \infty$, then this process is well-defined. The process y is stationary and ergodic with covariances

$$R(t, s) = \sum_{k=k(t,s)}^{\infty} G(t - s + k)G(k)^T = R(t - s) \tag{6.12}$$

where $k(t, s) = \max\{0, s - t\}$. The auxiliary variables ε act like inputs that generate the observed output process via a convolution relation, that is, via a linear, time invariant system. An essential difference with input-output convolution systems is that the auxiliary variables need not have an external significance. They merely facilitate the system description, and the process representation (6.11) is non-unique. The MA-representation (of the model) is called (causally) invertible if it can be rewritten as

$$\sum_{k=0}^{\infty} H(k)y(t - k) = \varepsilon(t) \tag{6.13}$$

with $\sum_{k=0}^{\infty} \|H(k)\| < \infty$. This is called an *autoregressive (AR)* representation. It expresses $y(t)$ in terms of its past values and an additional innovation $\varepsilon(t)$ which is uncorrelated with this past.

The processes described in Examples 6.2.2 and 6.2.3 are in a sense the building blocks of all stationary processes. The Wold decomposition theorem, which we state below, tells us that every stationary process can be decomposed into a moving average part and a perfectly predictable part.

Definition 6.2.3 A stationary process y is called *perfectly predictable* if there exists a prediction function with error zero, that is, for some function F there holds

$$E\|y(t) - F(y(s), s \leq t - 1)\|^2 = 0 \quad \forall t \in \mathbb{Z} \tag{6.14}$$

For example, a cyclical process as in Example 6.2.2 is perfectly predictable, in the case of Example 6.2.2 the function $F(y(s), s \leq t - 1)$ is given by $(2 \cos \omega)y(t - 1) - y(t - 2)$, as in (6.10).

A process is called *harmonic* if it is of the form $y(t) = \sum_{k=1}^{n} \alpha_k \sin(\omega_k t + \theta_k)$ with α_k, ω_k fixed and the θ_k independent and uniformly distributed on $[0, 2\pi)$. One can prove that such processes are stationary and perfectly predictable, also for $n \geq 2$ (cf. Example 6.2.2).

The proof of the following result requires mathematical methods that are beyond the scope of this book. The interested reader is referred to [10, page 187, Theorem 5.9.1].

Theorem 6.2.4 *Every stationary process y has a Wold decomposition $y = y_1 + y_2$ where*

(i) *y_1 and y_2 are stationary and uncorrelated,*
(ii) *y_1 is an invertible moving average process (6.11) with $\sum_{k=0}^{\infty} \|G(k)\|^2 < \infty$, and with $\sum_{k=0}^{\infty} \|H(k)\|^2 < \infty$ in the inverse representation (6.13),*
(iii) *y_2 is a perfectly predictable process with a linear function F in (6.14).*

Note that we only get here $\sum_{k=0}^{\infty} \|G(k)\|^2 < \infty$, and not the stronger condition $\sum_{k=0}^{\infty} \|G(k)\| < \infty$.

6.3　ARMA Processes

In this section we assume that any perfectly predictable component of the process has been removed, so that by Theorem 6.2.4 a moving average process remains. For the MA-process (6.11) the function $\hat{G}(z) = \sum_{k=0}^{\infty} G(k)z^{-k}$ is called the *filter* generating $y(t)$ from standard white noise. The relevance of filters is that the composition of processes corresponds with the multiplication of filters. Note its similarity with the transfer function of a linear time invariant system. The moving average process in principle involves an infinite number of parameters. This presents the problem to construct approximate models involving fewer parameters. A suitable approximation is a so-called *autoregressive moving average model*

$$y(t) = A_1 y(t-1) + \ldots + A_p y(t-p) + B_0 \varepsilon(t) + B_1 \varepsilon(t-1) + \ldots + B_q \varepsilon(t-q) \quad (6.15)$$

This is called an ARMA(p, q) model. In case $q = 0$ it is called an AR(p) model and for $p = 0$ it is called an MA(q) model. Here the filter $\hat{G}(z) = \sum_{k=0}^{\infty} G(k)z^{-k}$, is approximated by a rational function $A^{-1}(z)B(z)$, where $A(z) = I - A_1 z^{-1} - \ldots - A_p z^{-p}$ and $B(z) = B_0 + B_1 z^{-1} + \ldots + B_q z^{-q}$ are matrices with entries that are polynomials in z^{-1}. The next result states that these approximations become arbitrarily accurate if the orders p and q are chosen sufficiently large.

Theorem 6.3.1 *Every stationary process without perfectly predictable component can be approximated by AR and MA processes, that is, for every process y of the form (6.11) and for every $\delta > 0$ there exist orders p, q and an AR(p) process y_A and an MA(q) process y_M such that for all $t = 1, 2, \cdots$ we have $E \|y(t) - y_A(t)\|^2 < \delta$ and $E \|y(t) - y_M(t)\|^2 < \delta$.*

Proof　For simplicity we will only construct the approximation by an MA process in the zero mean, univariate case. The construction of the AR(p) process is more involved, and we refer to the literature for this.

So, let $y(t)$ be a scalar stationary process without perfectly predictable component. From Theorem 6.2.4 it follows that $y(t) = \sum_{k=0}^{\infty} G(k)\varepsilon(t-k)$, where $\sum_{k=0}^{\infty} G(k)^2 < \infty$

and with autoregressive representation (6.13) with the property that $\sum_{k=0}^{\infty} H(k)^2 < \infty$. By appropriate scaling we may assume that $G(0) = 1$, so that $H(0) = 1$ as well. The variance of the corresponding white noise process is denoted by $E(\varepsilon^2) = \sigma^2$.

For given $\delta > 0$ let q be such that $\sum_{k=q+1}^{\infty} G(k)^2 < \frac{\delta}{\sigma^2}$, then $y_M(t) := \sum_{k=1}^{q} G(k)\varepsilon(t - k)$ is an MA(q) process with the desired property. □

Although from an approximation viewpoint the AR and MA processes are sufficiently rich, ARMA models may provide a more accurate approximation with fewer parameters.

The moving average process (6.11) has an ARMA representation (6.15) if and only if the filter $\hat{G}(z) = \sum_{k=0}^{\infty} G(k)z^{-k}$ is a rational matrix function. However, the parameters of the model (6.15) are not uniquely determined. First, the noise process ε may be chosen in different ways. Second, for a given choice of ε, the factorization $\hat{G}(z) = A^{-1}(z)B(z)$ in terms of the polynomials $A(z)$ and $B(z)$ is not unique. We illustrate this by two simple examples.

Example 6.3.1 Consider the scalar MA(1) process y given by $y(t) = \varepsilon(t) + \theta\varepsilon(t-1)$ with $|\theta| < 1$, where ε is standard white noise. Then $\varepsilon(t) = \sum_{k=0}^{\infty}(-\theta)^k y(t - k)$, almost surely, which means that the process can be decomposed as $y(t) = \varepsilon(t) + f(y(s), s \leq t - 1)$. As $\varepsilon(t)$ is uncorrelated with the past observations of y, i.e. $\{y(s), s \leq t-1\}$, it follows that ε is the forward prediction error process corresponding to the process y. Actually, this is the Wold representation of Theorem 6.2.4(ii) for this process.

Now define the process ω by $\omega(t) = \sum_{k=0}^{\infty}(-\theta)^k y(t + 1 + k)$. Use $E(y(s)y(s-1)) = E(y(s-1)y(s)) = \theta$, $E(y(s)^2) = 1 + \theta^2$ and $Ey(s)y(t) = 0$ if $|s - t| \geq 2$ to check that ω is a standard white noise process and that $y(t) = \theta\omega(t) + \omega(t - 1)$. This is an alternative MA(1) representation of the process y. Actually, ω is the backward prediction error process corresponding to the process y. So the process y can be described by the filter $1 + \theta z^{-1}$ and also by the filter $\theta + z^{-1}$, by appropriate choice of the driving white noise process.

Example 6.3.2 Consider the bivariate MA(1) process

$$y_1(t) = \varepsilon_1(t) + \theta\varepsilon_2(t - 1),$$
$$y_2(t) = \varepsilon_2(t). \tag{6.16}$$

One easily derives the alternative AR(1) representation

$$y_1(t) - \theta y_2(t - 1) = \varepsilon_1(t)$$
$$y_2(t) = \varepsilon_2(t). \tag{6.17}$$

That is, this process can be written in ARMA form (6.15) with $A(z) = I$ and $B(z) = \begin{pmatrix} 1 & \theta z^{-1} \\ 0 & 1 \end{pmatrix}$ or alternatively with $A(z) = \begin{pmatrix} 1 & -\theta z^{-1} \\ 0 & 1 \end{pmatrix}$ and $B(z) = I$.

In practice one prefers representations with few parameters and where the noise process ε has a good interpretation. The first condition is related to the notion of coprimeness, the second to the notions of stationarity and invertibility.

Two matrices that are polynomial in z^{-1}, $A(z)$ and $B(z)$ are called *left coprime* if they have no non-trivial common left factors, i.e., if there are polynomial matrices C, A_1, B_1, in z^{-1}, such that $A(z) = C(z)A_1(z)$ and $B(z) = C(z)B_1(z)$, then C must be unimodular, i.e. $\det C(z)$ does not depend on z, and hence $C(z)^{-1}$ is also a polynomial in z^{-1}. In the univariate case, coprimeness means that the two polynomials A and B have no common factors. If A and B are not coprime, then one can find A_1 and B_1 such that $A_1^{-1}B_1 = A^{-1}B$, but (A_1, B_1) are of lower degree than (A, B).

Proposition 6.3.2 *If the stationary process y in (6.15) has no perfectly predictable component, then it can be represented by a coprime ARMA model.*

Proof For simplicity we only consider the univariate case, in the multivariate case the proof is more complicated.

Recall that multiplication of filters corresponds to composition of corresponding (MA or AR) representations. Write (6.15) for simplicity as $Ay = B\varepsilon$. Let F be the greatest common divisor of A and B, so that $A = F\tilde{A}$, $B = F\tilde{B}$, and $\tilde{A}$ and $\tilde{B}$ have no common factors. Introduce $x = \tilde{A}y - \tilde{B}\varepsilon$. Define the processes y_1 and y_2 by $\tilde{A}y_1 = \tilde{B}\varepsilon$ and $\tilde{A}y_2 = x$. Then $y = y_1 + y_2$ because $A(y_1 + y_2) = B\varepsilon$. We have $Ay_2 = F\tilde{A}y_2 = Fx = F(\tilde{A}y - \tilde{B}\varepsilon) = 0$. This means that y_2 is perfectly predictable and hence, by assumption $y_2 = 0$. Therefore $y = y_1$ and $\tilde{A}y = \tilde{B}\epsilon$ is a coprime ARMA representation.   □

Further, the ARMA model (6.15) is said to be *stationary* if the process it represents can be written as in (6.11) with $\sum \|G(k)\| < \infty$, and the model is called *invertible* if the process it represents can be written as in (6.13) with $\sum \|H(k)\| < \infty$. The corresponding filters are called *causal*, and *invertible*, respectively. So stationarity of the model means that y is related to ε in a causal way, invertibility means that ε is related to y in a causal way. These conditions mean that the process ε consists of the (forward) prediction errors of the process y. The first filter in Example 6.3.1 is causal with a causal inverse. A filter $\hat{G}$ is called *anticausal* if it can be represented as $\sum_{k=-\infty}^{0} G(k)z^{-k}$ with $\sum_{k=-\infty}^{0} \|G(k)\| < \infty$. The second filter in Example 6.3.1 is causal with an anticausal inverse.

Theorem 6.3.3 *Consider a stationary process with coprime ARMA representation (6.15).*

(i) *The representation is stationary if and only if* det $A(z)$ *has all its roots inside the unit circle* $|z| < 1$; *in this case (6.11) is obtained by the filter* $\hat{G}(z) = A(z)^{-1}B(z)$.

(ii) *The representation is invertible if and only if* det $B(z)$ *has all its roots inside the unit circle, and (6.13) is then given by the filter* $\hat{H}(z) = B(z)^{-1}A(z)$.

Proof   Again, for simplicity we only consider the univariate case. The multivariate result follows in a similar way by using the Smith form of polynomial matrices (see, e.g., [19]).

First assume that A has all its roots inside the unit circle. Then $\hat{G}(z) = A^{-1}(z)B(z)$ is a rational function which has all its poles inside the unit disk and has limit B_0 for $z \to \infty$. It is well known in scalar complex function theory that such a rational function $\hat{G}(z)$, has a series expansion $\sum_{k=0}^{\infty} G(k)z^{-k}$ with $\sum_{k=0}^{\infty} |G(k)| < \infty$.

For the converse we first remark that $A(z)$ and $B(z)$ do not have a common zero because they are coprime. Hence any zero of $A(z)$ gives a pole of $\hat{G}(z) = A^{-1}(z)B(z)$. Since $\hat{G}(z)$, has a series expansion $\sum_{k=0}^{\infty} G(k)z^{-k}$ with $\sum_{k=0}^{\infty} |G(k)| < \infty$ it is known from complex function theory that $\hat{G}(z)$ has all its poles inside the unit disk. Thus $A(z)$ has all its zeros inside the unit disk.

Part (ii) is proved in a similar way.   □

Let y be a stationary process without perfectly predictable component and with a coprime ARMA representation (6.15). In the next chapter we will show that this process then also has a stationary and invertible ARMA representation. We conclude that within this setting it is no restriction to assume that an ARMA model is stationary and invertible. However, this representation is still not unique. For instance, the representation in (6.16) and (6.17) are both coprime, stationary and invertible. To obtain uniqueness one should impose additional restriction on the parameters in ARMA models.

6.4   State Space Models

The state of a system contains all the information on the interdependence between the past and the future of the system. Given the current state, the future evolution becomes independent from the past.

Consider stochastic vectors x and y. The best linear prediction of x in terms of y is a stochastic vector Ly, where L is a matrix such that $E(x - Ly)(x - Ly)^T$ is minimal. The solution is then given by projecting each component $x_1, \ldots, x_n$ of x orthogonally on the space spanned by the components $y_1, \ldots, y_m$ of y. This means that we look for row vectors l_j, $j = 1, \ldots, n$, such that $E(x_j - l_j y)y_k = 0$ for $k = 1, \ldots, m$. This gives m linear equations for the l_j, which can be written as $Ex_j y^T = l_j Eyy^T$. In the case that Eyy^T is invertible we find $l_j = Ex_j y^T (Eyy^T)^{-1}$. For the matrix L we therefore get $L = Exy^T (Eyy^T)^{-1}$. For this Ly we use the notation $E(x|y)$ and call this best linear prediction of x based on y, or the conditional expectation of x based on y. Therefore, $E(x|y) = 0$ if and only if $E(xy^T) = 0$, so that all components of x and y are uncorrelated. In a similar way we define $E(x|y(1), y(2), \ldots)$ as the best linear approximation of x by a combination of the components of the vectors $y(1), y(2), \ldots$. This then has the form $\Sigma_j H(j)y(j)$ for appropriate matrices $H(j)$.

Definition 6.4.1 The process x is called a *state process* for the process y if for every $t \in \mathbb{Z}$

$$
\begin{aligned}
E(y(t+k)|x(t), y(t-1), y(t-2), \ldots) &= E(y(t+k)|x(t)) \qquad \forall k \geq 0, \\
E(y(t-k)|x(t), y(t), y(t+1), \ldots) &= E(y(t-k)|x(t)) \qquad \forall k \geq 1.
\end{aligned}
\tag{6.18}
$$

Thus the process x summarizes all correlations between the past and the future of the process y.

We consider models of the form

$$
\begin{cases}
x(t+1) &= Ax(t) + \varepsilon_1(t), \qquad t \in \mathbb{Z} \\
y(t) &= Cx(t) + \varepsilon_2(t),
\end{cases}
\tag{6.19}
$$

where $(\varepsilon_1, \varepsilon_2)$ is a joint white noise process with covariance matrix

$$
E \begin{pmatrix} \varepsilon_1(t) \\ \varepsilon_2(t) \end{pmatrix} \begin{pmatrix} \varepsilon_1(t) \\ \varepsilon_2(t) \end{pmatrix}^T = \begin{pmatrix} \Sigma_{11} & \Sigma_{12} \\ \Sigma_{21} & \Sigma_{22} \end{pmatrix} = \Sigma.
\tag{6.20}
$$

We restrict the attention to stationary representations, where the filter generating x from ε_1 is causal. According to Theorem 6.3.3 (i) this means that $\det(I - Az^{-1})$ has all its zeros in the unit disk. This is equivalent to A being a stable matrix, that is, A has all its eigenvalues in the open unit disc. Further we assume for simplicity that all processes considered have zero mean.

Proposition 6.4.2 *If the matrix A is stable, then the process x in (6.19) is a Markov process, that is,*

$$
E(x(t+1)|x(s), s \leq t) = E(x(t+1)|x(t)),
$$

and it is a state for the process y (6.19).

Proof   Since A is stable we have $x(t) = \sum_{k=1}^{\infty} A^{k-1}\varepsilon_1(t-k)$. If $s \leq t$ then $E(x(s)\varepsilon_1(t)^T) = \sum_{k=1}^{\infty} A^{k-1}E(\varepsilon_1(s-k)\varepsilon_1(t)^T) = 0$. We conclude that $\varepsilon_1(t)$ and $x(s)$ are uncorrelated. Therefore

$$E(x(t+1)|x(s), s \leq t) = E(Ax(t) + \varepsilon_1(t)|x(s), s \leq t) = E(Ax(t)|x(s), s \leq t).$$

Now by definition $E(Ax(t)|x(s), s \leq t) = Ax(t) = E(Ax(t)|x(t))$. Again using that $\varepsilon_1(t)$ and $x(t)$ are uncorrelated we get

$$E(Ax(t)|x(t)) = E(Ax(t) + \varepsilon_1(t)|x(t)) = E(x(t+1)|x(t)).$$

Further, $y(t) = \varepsilon_2(t) + \sum_{k=1}^{\infty} CA^{k-1}\varepsilon_1(t-k)$. This shows that $(\varepsilon_1(t), \varepsilon_2(t))$ and $y(s), s \leq t-1$, are uncorrelated. For $v \geq t$ there holds $y(v) = CA^{v-t}x(t) + \varepsilon_2(v) + \sum_{k=1}^{v-t} CA^{k-1}\varepsilon_1(v-k)$, so that

$$E(y(v)|x(t), y(s), s \leq t-1) = CA^{v-t}x(t) = E(y(v)|x(t)).$$

This proves one part of (6.18).

Because $y(v)$ is for $v \geq t$ a linear expression in $x(t)$, $\varepsilon_2(v)$, and $\varepsilon_1(v)$, we get

$$E(y(t-k)|x(t), y(t), y(t+1), \ldots) = E(y(t-k)|x(t), \varepsilon_1(t), \varepsilon_2(t), \varepsilon_1(t+1), \varepsilon_2(t+1), \ldots).$$

Once again use that $y(t-k)$ and $\varepsilon_1(t+j)$, $\varepsilon_2(t+j)$, $j \geq 0$, are uncorrelated. We get $E(y(t-k)|x(t), y(t), y(t+1), \ldots) = E(y(t-k)|x(t))$. $\square$

The state space model (6.19) is of much practical use because of its simple first order structure. We will now show which processes can be represented in state space form and how such representations can be obtained. The following result is the stochastic analogue of Theorem 2.3.3 for deterministic input-output systems. Recall the a model is stationary means that the corresponding filter is causal.

Theorem 6.4.3 *A stationary process y can be represented in state space form* (6.19) *with a stable matrix A if and only if it can be represented by a stationary ARMA model* (6.15).

Proof   First suppose that the model (6.15) is given. For the moment denote $\hat{A}(z) = I - A_1 z^{-1} - \cdots - A_p z^{-p}$. Let $m := \max\{p, q\}$ and define $A_i = 0, i \geq p+1$, and $B_j = 0$, $j \geq q+1$. Let $x(t) = (x_1(t)^T, \ldots, x_m(t)^T)^T$ where

$$x_i(t) := \sum_{k=1}^{m} A_{k+i-1}y(t-k) + \sum_{k=1}^{m} B_{k+i-1}\varepsilon(t-k), \quad i = 1, \ldots, m.$$

This gives a representation (6.19) with $C = \begin{pmatrix} I & 0 & \dots & 0 \end{pmatrix}$, $\varepsilon_2(t) = B_0\varepsilon(t)$ and $\varepsilon_1(t) = B\varepsilon(t)$, where

$$A = \begin{pmatrix} A_1 & I & 0 & \dots & 0 \\ A_2 & 0 & I & & 0 \\ \vdots & \vdots & \vdots & \ddots & \vdots \\ A_{m-1} & 0 & 0 & \dots & I \\ A_m & 0 & 0 & \dots & 0 \end{pmatrix}, \qquad B = \begin{pmatrix} A_1 B_0 + B_1 \\ A_2 B_0 + B_2 \\ \vdots \\ A_{m-1} B_0 + B_{m-1} \\ A_m B_0 + B_m \end{pmatrix}.$$

It follows that $\det(zI - A) = \det z^m \hat{A}(z)$, and thus by Theorem 6.3.3 the stationary ARMA model gives a stable matrix A.

Conversely, for a given model (6.19) with A a stable $n \times n$ matrix, define $\hat{P}(z) := I - \frac{1}{z}A$ and $\hat{p}(z) := \det \hat{P}(z)$. Further let the polynomial matrix $\hat{P}^+(z)$ in $\frac{1}{z}$ denote the adjoint of $\hat{P}(z)$, so that $\hat{P}^+(z)\hat{P}(z) = \hat{p}(z) \cdot I$. Like in the proof of Theorem 6.3.3 we relate to the matrix polynomials $\hat{P}$, $\hat{p} \cdot I$, and $\hat{P}^+$ the corresponding filters P, p, and P^+. This way the state process in (6.19) can be written as $Px = \varepsilon_1$, so that $py = Cpx + p\varepsilon_2 = CP^+Px + p\varepsilon_2 = CP^+\varepsilon_1 + p\varepsilon_2$. This is an ARMA$(n, n)$ representation with AR polynomial $\hat{p}(z) \cdot I$. This representation is stationary because $\hat{p}(z)$ has its roots inside the unit circle.   □

In the univariate case the construction above leads to a minimal state space representation if $A_p \neq 0$ and $B_q \neq 0$ in (6.15). In the multivariate case it may be non-minimal.

Stochastic realization theory concerns the relationship between stochastic processes and their state space representations. Let y be a given stationary process with representation (6.19) where A is a stable matrix. We assume that Σ_{22} in (6.20) is invertible, so that y has no perfectly predictable component. The autocovariances of the process are easily obtained from this representation.

Proposition 6.4.4 *Let A be a stable matrix. Then the autocovariances of the process y in (6.19), (6.20) with Σ_{22} invertible, are given by*

$$R(0) = C\Pi C^T + \Sigma_{22}, \tag{6.21}$$

$$R(k) = CA^{k-1}M, \quad k \geq 1, \tag{6.22}$$

where $M = E(x(t+1)y(t)^T) = A\Pi C^T + \Sigma_{12}$, and $\Pi = E(x(t)x(t)^T)$ satisfies the Stein equation

$$\Pi = A\Pi A^T + \Sigma_{11}. \tag{6.23}$$

Proof As A is stable the state process (6.19) satisfies $x(s) = \sum_{k=1}^{\infty} A^{k-1}\varepsilon_1(s-k)$, so that $\varepsilon_1(t)$ and $\varepsilon_2(t)$ are uncorrelated with $x(s)$ for $s \leq t$. Further, $x(t)$ is a stationary process, so that $E(x(t)x(t)^T) = E(x(t+1)x(t+1)^T)$. Hence

$$\Pi = E(Ax(t) + \varepsilon_1(t))(Ax(t) + \varepsilon_1(t))^T = A\Pi A^T + \Sigma_{11}.$$

In addition,

$$R(0) = E(y(t)y(t)^T) = E(Cx(t) + \varepsilon_2(t))(Cx(t) + \varepsilon_2(t))^T = C\Pi C^T + \Sigma_{22}$$

and

$$M = E(x(t+1)y(t)^T) = E(Ax(t) + \varepsilon_1(t))(Cx(t) + \varepsilon_2(t))^T = A\Pi C^T + \Sigma_{12}.$$

Furthermore, for $k \geq 1$ there holds $x(t) = A^{k-1}x(t-k+1) + \sum_{j=1}^{k-1} A^{j-1}\varepsilon_1(t-j)$, so that

$$\begin{aligned} R(k) = E\{y(t)y(t-k)^T\} &= \\ &= E\{[CA^{k-1}x(t-k+1) + \sum_{j=1}^{k-1} CA^{j-1}\varepsilon_1(t-j)]y(t-k)^T\} \\ &= CA^{k-1}M \end{aligned}$$

as $\varepsilon_1(t)$ is uncorrelated with $y(s)$ for $s < t$. □

Next we consider the converse problem of weak stochastic realization. In this case the autocovariances $R(k)$ (with $\sum \|R(k)\| < \infty$) of the process are given and the problem is to determine a state space model (6.19), (6.20) with the same autocovariances. The foregoing result expresses the involved restrictions on the parameters in this model. The first step in the solution of the weak stochastic realization problem is to construct the matrices (A, C, M) so that $R(k) = CA^{k-1}M$, $k \geq 1$. This may be done using for example the algorithm of Sect. 3.4. The noise covariance matrices in (6.20) can then be derived from the state covariance matrix $\Pi = E(x(t)x(t)^T)$ because

$$\Sigma = \begin{pmatrix} \Sigma_{11} & \Sigma_{12} \\ \Sigma_{21} & \Sigma_{22} \end{pmatrix} = \begin{pmatrix} \Pi - A\Pi A^T & M - A\Pi C^T \\ M^T - C\Pi A^T & R(0) - C\Pi C^T \end{pmatrix}. \tag{6.24}$$

Therefore it remains to determine a positive semidefinite matrix Π such that Σ is positive semidefinite. The next theorem describes the solution set in qualitative terms. For a proof we refer to [12].

Theorem 6.4.5 *For given $R(0)$ and minimal (A, C, M), the set of positive semidefinite matrices Π such that (6,24) is positive semidefinite is convex and bounded. It has a minimal solution Π_- and a maximal solution Π_+ such that for all other solutions $\Pi_- \leq \Pi \leq \Pi_+$.*

A realization (6.19) of the covariances $\{R(k), k \in \mathbb{Z}\}$ is called *minimal* if both the number of state variables $x(t)$ and the number of independent noise variables in $\begin{pmatrix} \varepsilon_1(t) \\ \varepsilon_2(t) \end{pmatrix}$ are as small as possible. The minimal number of states is obtained if (A, C, M) is a minimal triple. The number of independent noise terms is minimized by selecting a matrix Π of the solution set in Theorem 6.4.5 that minimizes the rank of Σ in (6.24). As Σ_{22} is assumed to be invertible, it follows that

$$\Sigma = \begin{pmatrix} \Sigma_{11} & \Sigma_{12} \\ \Sigma_{21} & \Sigma_{22} \end{pmatrix} = \begin{pmatrix} I & \Sigma_{12} \\ 0 & \Sigma_{22} \end{pmatrix} \begin{pmatrix} Z & 0 \\ 0 & \Sigma_{22}^{-1} \end{pmatrix} \begin{pmatrix} I & 0 \\ \Sigma_{21} & \Sigma_{22} \end{pmatrix},$$

where $Z = \Sigma_{11} - \Sigma_{12}\Sigma_{22}^{-1}\Sigma_{21}$. The rank is minimized by taking $Z = 0$, so that Π satisfies the algebraic Riccati equation

$$\Pi = A\Pi A^T + (M - A\Pi C^T)(R(0) - C\Pi C^T)^{-1}(M - A\Pi C^T)^T. \tag{6.25}$$

The solution of this equation is in general not unique. Of particular interest is the minimal solution Π_-, which is obtained as the limit for $k \to \infty$ of the recursion

$$\Pi_0 = 0, \quad \Pi_{k+1} = A\Pi_k A^T + (M - A\Pi_k C^T)(R(0) - C\Pi_k C^T)^{-1}(M - A\Pi_k C^T)^T.$$

For this solution the state covariance matrix is as small as possible.

6.5   Spectra and the Frequency Domain

Let y be a stationary process with zero mean and covariances

$$R(k) = Ey(t)y(t-k)^T, k \in \mathbb{Z}. \tag{6.26}$$

Definition 6.5.1 The *spectrum* of a stationary process is defined by the formal power series

$$S(z) = \frac{1}{2\pi} \sum_{k=-\infty}^{\infty} R(k)z^{-k}. \tag{6.27}$$

The spectrum is a well-defined function of the complex variable $z = e^{i\omega}$ on the unit circle if we impose the condition

$$\sum_{k=-\infty}^{\infty} ||R(k)|| < \infty. \tag{6.28}$$

As $S(e^{i\omega}) = S(e^{-i\omega})^T$ it suffices to consider the spectrum only for $\omega \in [0, \pi]$. For each ω, the value of $S(e^{i\omega})$ is a complex-valued positive semidefinite matrix, and in the scalar case $S(e^{i\omega})$ is real-valued and non-negative. In Fourier Analysis the following theorem is well known.

Theorem 6.5.2 *The autocovariances of a process with spectrum S, which satisfies the condition $\int_{-\pi}^{\pi} ||S(e^{i\omega})|| d\omega < \infty$, are given by*

$$R(k) = \int_{-\pi}^{\pi} e^{ik\omega} S(e^{i\omega}) d\omega \tag{6.29}$$

and (6.28) is satisfied.

Note that $\int_{-\pi}^{\pi} e^{in\omega} d\omega = 0$ if $n \neq 0$ and that this integral equals 2π if $n = 0$. So if (6.28) is satisfied, then

$$\int_{-\pi}^{\pi} e^{ik\omega} S(e^{i\omega}) d\omega = \sum_{l=-\infty}^{\infty} \frac{1}{2\pi} R(l) \int_{-\pi}^{\pi} e^{i(k-l)\omega} d\omega = R(k).$$

The formulas (6.27) and (6.29) show that the covariance sequence $\{R(k), k \in \mathbb{Z}\}$ and the function $S(e^{i\omega})$ contain the same information.

The condition (6.28) on the summability of the norms of the autocovariances is a necessary one for the spectrum to be a well-defined function. To see this consider the spectrum of a cyclic process $y(t) = \sin(\omega_0 t + \theta)$. Then we know from Example 6.2.2 that $R(k) = \frac{1}{2} \cos(\omega_0 k)$. Hence the spectrum

$$S(e^{i\omega}) = \frac{1}{4\pi} \left(\sum_{k=-\infty}^{\infty} \cos(\omega k) \cos(\omega_0 k) + i \sum_{k=-\infty}^{\infty} \sin(\omega k) \cos(\omega_0 k) \right).$$

As $\sin(\omega(-k)) = -\sin(\omega k)$ the second summand vanishes, and since $\cos(\omega(-k)) = \cos(\omega k)$ we have

$$S(e^{i\omega}) = \frac{1}{4\pi} \left(1 + 2 \sum_{k=1}^{\infty} \cos(\omega k) \cos(\omega_0 k) \right).$$

Now from Fourier Analysis it is known that the sum on the right will converge to zero if $\omega \neq \omega_0$ and to ∞ when $\omega = \omega_0$. So the limit is a Dirac-delta function. To get an impression, the graph below shows the sum of the first 40 and the first 100 terms in the series for $\omega_0 = \frac{\pi}{4}$, and the second graph shows 1024 terms. Note the different scales on the y-axis.

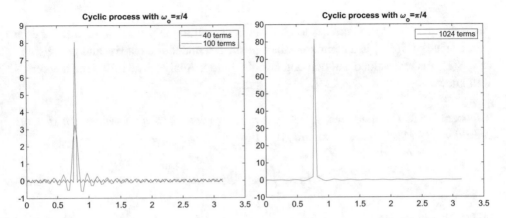

We now consider the spectrum of a moving average process

$$y(t) = \sum_{k=0}^{\infty} G(k)\varepsilon(t - k) \tag{6.30}$$

where ε is a white noise process with $E\varepsilon(t)\varepsilon(t)^T = I$, and where $\sum_{k=0}^{\infty} \|G(k)\| < \infty$.

Theorem 6.5.3 *The spectrum of the moving average process* (6.30) *is given by*

$$S(z) = \frac{1}{2\pi}\hat{G}(z)\hat{G}^T(z^{-1}) \tag{6.31}$$

where $\widehat{G}(z) := \sum_{k=0}^{\infty} G(k)z^{-k}$.

Proof   It follows from (6.30) that

$$R(k) = Ey(t)y^T(t - k) = \sum_{i=0}^{\infty} G(i + k)G^T(i),$$

which is also the coefficient of z^{-k} in

$$\widehat{G}(z)\widehat{G}^T(z^{-1}) = \sum_{i=0}^{\infty}\sum_{j=0}^{\infty} G(j)G^T(i)z^{i-j}.$$

These series are well-defined as $\sum_{k=0}^{\infty}\|G(k)\| < \infty$.   □

From this result we easily obtain the spectrum of an ARMA process with a coprime, stationary and invertible representation

$$y(t) = A_1 y(t-1) + \ldots + A_p y(t-p) + B_0\varepsilon(t) + B_1\varepsilon(t-1) + \ldots + B_q\varepsilon(t-q). \quad (6.32)$$

Theorem 6.5.4 *The spectrum of an ARMA process (6.32), where ε is a standard white noise, is given by*

$$S(z) = \frac{1}{2\pi} A^{-1}(z)B(z)B^T(z^{-1})(A^T)^{-1}(z^{-1}). \quad (6.33)$$

Proof According to Theorem 6.3.3 the filter of the moving average representation of the process is given by $\widehat{G}(z) = A^{-1}(z)B(z)$. The result then follows from Theorem 6.5.3. □

Recall that a stochastic system has a finite dimensional state space realization if and only its filter is rational. We proved that if the filter is rational, then the spectrum is rational. Conversely, if the spectrum is rational, one can prove that the filter is also rational. So, stochastic systems have a finite dimensional state space realization if and only if the spectrum is rational. This is similar to the result for deterministic input-output systems in terms of the transfer function.

Example 6.5.1 For univariate ARMA processes with $E\{\varepsilon(t)\}^2 = \sigma_\varepsilon^2$ we obtain

$$S(e^{i\omega}) = \frac{\sigma_\varepsilon^2}{2\pi}\frac{b(e^{-i\omega})b(e^{i\omega})}{a(e^{-i\omega})a(e^{i\omega})}.$$

Here a and b are scalar polynomials.

In particular, a standard white noise process has spectrum $S(e^{i\omega}) = (2\pi)^{-1}$, so that the spectrum is constant for all frequencies. The filter of an AR(1) process $y(t) = \alpha y(t-1) + \varepsilon(t)$ with $|\alpha| < 1$ is given by

$$\widehat{G}(z) = A(z)^{-1} = \frac{1}{1-\alpha z^{-1}} = \sum_{k=0}^{\infty}\alpha^k z^{-k}.$$

Hence the autocorrelations are $R(k) = \sum_{i=0}^{\infty} \alpha^{k+i} \alpha^i = \alpha^k/(1 - \alpha^2)$, $k \geq 0$. The spectrum of this process is given by

$$S(e^{i\omega}) = \frac{1}{2\pi(1 + \alpha^2 - 2\alpha \cos \omega)}.$$

If $\alpha > 0$ the spectrum is monotonically decreasing in $\omega \in [0, \pi)$, indicating that the lower frequencies are dominant in this process. This is in agreement with the positive correlations $R(k) = \alpha^k/(1 - \alpha^2)$, $k \geq 0$, so that this process tends to oscillate less as compared with white noise. On the other hand, if $\alpha < 0$ the process oscillates more heavily, and this translates into a monotonically increasing spectrum in $\omega \in [0, \pi)$. So here the high frequency components are dominant. Such interpretations of the spectrum can often be made for stationary processes.

Finally, we consider the spectrum of the state space model (6.19). Proposition 6.4.4 describes the $R(k)$'s through formulas (6.21) and (6.22). Thus, let A be stable, and let the process $y(t)$ be given by (6.19). Then

$$S(z) = \frac{1}{2\pi} \left(\Sigma_{22} + C\Pi C^T + \sum_{k=1}^{\infty} CA^{k-1} M z^{-k} + \sum_{k=1}^{\infty} M^T (A^T)^{k-1} C^T z^k \right),$$

where $M = \Sigma_{12} + A\Pi C^T$ and Π is the unique solution to $\Pi - A\Pi A^T = \Sigma_{11}$. Now

$$\sum_{k=1}^{\infty} CA^{k-1} z^{-k} M = C(zI - A)^{-1} M$$

and

$$\sum_{k=1}^{\infty} M^T (A^T)^{k-1} C^T z^k = z M^T (I - zA)^{-1} C^T.$$

Hence we arrive at the following theorem.

Theorem 6.5.5 *The spectrum of a stationary process $y(t)$ which is given by (6.19), where A is a stable matrix, and where $(\varepsilon_1, \varepsilon_2)$ is a joint white noise process with covariance matrix (6.20), is given by*

$$S(z) = \frac{1}{2\pi} \left(\Sigma_{22} + C\Pi C^T + C(zI - A)^{-1} M + z M^T (I - zA)^{-1} C^T \right), \tag{6.34}$$

where $M = \Sigma_{12} + A\Pi C^T$ and Π is the unique solution to $\Pi - A\Pi A^T = \Sigma_{11}$.

6.6 Stochastic Input-Output Systems

In this section we present a short description of the broad class of stochastic input-output systems that can be described by the convolutions

$$y(t) = \sum_{k=0}^{\infty} G_1(k)u(t-k) + \sum_{k=0}^{\infty} G_2(k)\varepsilon(t-k). \tag{6.35}$$

In this case the output process y is generated by a linear, time invariant system driven by control inputs u and an auxiliary white noise process ε. If $\widehat{G}_1(z) = \sum_{k=0}^{\infty} G_1(k)z^{-k}$ and $\widehat{G}_2(z) = \sum_{k=0}^{\infty} G_2(k)z^{-k}$ are both rational, so that $\widehat{G}_i(z) = \widehat{A}_i(z)^{-1}\widehat{B}_i(z)$ for polynomial matrices $\widehat{A}_i, \widehat{B}_i, i = 1, 2$, then the process y can be represented in ARMAX form (6.4). Indeed, define $A = (\det \widehat{A}_1)\widehat{A}_2$, $B = (\det \widehat{A}_1)\widehat{A}_2\widehat{A}_1^{-1}\widehat{B}_1$, and $C = (\det \widehat{A}_1)\widehat{B}_2$, then A, B and C are polynomial matrices and (6.35) can be written as

$$\begin{aligned} y(t) + A_1 y(t-1) + \ldots + A_L y(t-L) &= B_0\varepsilon(t) + B_1\varepsilon(t-1) + \ldots \\ &+ B_L\varepsilon(t-L) + C_0 u(t) + C_1 u(t-1) + \ldots + \ldots C_L u(t-L), \end{aligned} \tag{6.36}$$

where $\widehat{A}(z) = I - \sum_{k=1}^{L} A_k z^{-k}$, $\widehat{B}(z) = \sum_{k=0}^{L} B_k z^{-k}$, and $\widehat{C}(z) = \sum_{k=0}^{L} C_k z^{-k}$. The output process y of (6.35) can be decomposed as $y = y_1 + y_2$, where the process $y_1(t) = \sum_{k=0}^{\infty} G_1(k)u(t-k)$ describes the impact of the control inputs and the process $y_2(t) = \sum_{k=0}^{\infty} G_2(k)\varepsilon(t-k)$ the impact of the disturbance terms. If $\widehat{G}_1$ and $\widehat{G}_2$ are both rational, then y_1 has a finite dimensional deterministic realization and y_2 a finite dimensional stochastic realization. By combining these models we obtain a state space representation of the ARMAX process of the form

$$\begin{aligned} x(t+1) &= Ax(t) + Bu(t) + E\varepsilon(t) \\ y(t) &= Cx(t) + Du(t) + F\varepsilon(t) \end{aligned} \tag{6.37}$$

The models (6.35), (6.36) and (6.37) describe the stochastic properties of the output process y only after the interdependence between u and ε has been specified. One possible interpretation of these models is that they describe the stochastic evolution of y conditionally, for a fixed input trajectory u. This is the so-called open-loop interpretation. A similar interpretation is possible if the processes u and ε are uncorrelated. In many cases, however, the input will be correlated with the noise. This is the case, for example, in controlled processes where u depends on past observations of y. This leads to a closed-loop interpretation. For example, a finite dimensional controller could be of the form $\sum_{k=0}^{N} P(k)u(t-k) = \sum_{k=0}^{N} Q(k)y(t-k) + \sum_{k=0}^{N} R(k)\eta(t-k)$, where η is a white noise process modelling additional influences on the control input. Combining this with (6.36) would lead to a joint stochastic ARMA model for the external variables (u, y).

Filtering and Prediction

7

Stochastic systems can be applied for forecasting purposes. The classical solution for filtering, smoothing and prediction of linear systems was proposed by Wiener and Kolmogorov in terms of spectral representations. The Kalman filter is a much more efficient, recursive solution in terms of state space models.

7.1 The Filtering Problem

For a deterministic input-output system the future outputs are exactly known once the future inputs have been chosen. For stochastic systems, however, the future disturbances are unknown, and therefore the future outputs can only be predicted with some error. The objective is to construct predictions that minimize the prediction error in some sense. Forecasting is one of the major applications of stochastic systems, in economics, engineering and many other disciplines.

The filtering problem is formalized as follows. Suppose that two jointly stationary processes, y and z, are mutually correlated and that the covariances (or the spectrum) of the joint process are completely known, but that only y is observed and z is not. As an example, you may think of z as the state in a model of the type (6.19), and y as the output. The aim is to form an optimal reconstruction $\hat{z}$ of the unobserved process z on the basis of the observed process y, via some function f of (possibly only some of) the values $\{y(s); s \in \mathbb{Z}\}$, i.e., $f\{y(s); s \in \mathbb{Z}\}$. So the problem will be to determine this function f.

If for the reconstruction $\hat{z}$ of $z(t)$ only the past and current values of y, i.e. $\{y(s); s \leq t\}$, can be used, this is called *filtering*. If only $\{y(s); s \leq t - m\}$ for some $m > 0$ can be used this is the *m-step ahead prediction problem*, and if $m < 0$ this is called *smoothing*. The case where $m = -\infty$ is called *unrestricted smoothing*. The one-step ahead prediction problem is often called *the filtering problem*, and we will pay special attention to this case. For this

© The Author(s), under exclusive license to Springer Nature Switzerland AG 2021
C. Heij et al., *Introduction to Mathematical Systems Theory*,
https://doi.org/10.1007/978-3-030-59654-5_7

case, as objective we consider here the minimization of the mean squared prediction error,

$$E \| z(t) - f\{y(s); s \le t - 1\} \|^2 \tag{7.1}$$

In particular, if $z = y$ then this corresponds to the *one-step ahead prediction* of a process based on the past observed values of this process.

The next proposition states that for the most common instances of the problem, when the process $\{y, z\}$ is jointly Gaussian, we may as well assume that the function f is a linear function.

Proposition 7.1.1 *The optimal predictor in* (7.1) *is given by the conditional expectation* $E\{z(t) \mid y(s); s \le t - 1\}$. *This is a linear function in case the process* $\{y, z\}$ *is jointly Gaussian.*

Moreover, if the prediction function f in (7.1) *is restricted to be linear, then the optimal solution for arbitrary distributions is as in the Gaussian case.*

We do not give a detailed proof here, for a proof see, e.g., [10]. The outline of the proof is as follows. The optimal solution in (7.1) is obtained by projecting the components of $z(t)$ on the space of all (measurable) functions spanned by the components of $\{y(s); s \le t - 1\}$. This projection is the definition of conditional expectation. In the Gaussian case the conditional expectation is a linear function. If we require that the predictor in (7.1) is linear, then the prediction error criterion depends only on the first and second moments of the processes, so the optimal solution is the same as in the Gaussian case.

For background material on conditional expectation, see, e.g., [64, Chapter 7, Sections 6 and 7].

Restricting the attention to linear predictors, in particular, to the Gaussian case, the filtering problem consists of determining the parameters of an optimal predictor. As the processes y and z are assumed to be jointly stationary, by time-invariance, instances of the filtering problem all reduce to finding matrices $F(k)$ such that

$$\hat{z}(t) = \sum_{k=-\infty}^{\infty} F(k) y(t - k). \tag{7.2}$$

In the m-step ahead problem the filter F satisfies $F(k) = 0$ for $k \le m - 1$. In particular, the optimal one-step ahead predictor $\hat{z}(t)$ is given by a time invariant filter of the form

$$\hat{z}(t) = \sum_{k=1}^{\infty} F(k) y(t - k). \tag{7.3}$$

For the smoothing problem we have

$$\hat{z}(t) = \sum_{k=m}^{\infty} F(k)y(t-k)$$

for some $m < 0$. The case $m = -\infty$ is called the *unrestricted smoothing problem*.

Obviously, the problem is now to determine the matrices $F(k)$ based on given data for the processes y and z. The solution depends essentially on what we assume to be known.

As a first illustration, let y be a purely nondeterministic process with invertible moving average representation $y(t) = \sum_{k=0}^{\infty} G(k)\varepsilon(t-k)$. Here the process ε is standard white noise and can be written as $\varepsilon(t) = \sum_{k=0}^{\infty} H(k)y(t-k)$. We denote the optimal linear m-step ahead predictor of y on the basis of its own past by $\hat{y}(t+m \mid t) = E(y(t+m) \mid y(s), s \leq t)$. We assume here that the matrices $G(k)$ are known, and we are interested in the case $y = z$. In this case the m-step ahead prediction problem has a particularly easy solution.

Proposition 7.1.2 *Let* $y(t) = \sum_{k=0}^{\infty} G(k)\varepsilon(t-k)$ *be a causal and invertible MA representation with ε standard white noise. Then the optimal linear m-step ahead predictor of y is given by*

$$\hat{y}(t+m \mid t) = \sum_{k=m}^{\infty} G(k)\varepsilon(t+m-k) \tag{7.4}$$

and the covariance matrix of the prediction error $\hat{y}(t+m|t) - y(t+m)$ is equal to $\sum_{k=0}^{m-1} G(k)G(k)^T$.

Proof   Observe that $y(t+m) = \sum_{k=0}^{m-1} G(k)\varepsilon(t+m-k) + \sum_{k=m}^{\infty} G(k)\varepsilon(t+m-k)$. Now $\hat{y}(t+m|t) = \sum_{k=m}^{\infty} G(k)\varepsilon(t+m-k)$, because for $s > t$ the white noise $\varepsilon(s)$ is uncorrelated with $y(t)$, and due to the invertibility of the moving average representation

$$E(y(t+m)|y(s); s \leq t) = E(y(t+m)|\varepsilon(s); s \leq t).$$

This proves (7.4) and also

$$y(t+m) - \hat{y}(t+m|t) = \sum_{k=0}^{m-1} G(k)\varepsilon(t+m-k).$$

The latter formula gives that the error that it expresses has covariance matrix $\sum_{k=0}^{m-1} G(k)G(k)^T$.   □

Let us consider what this means for a causal and invertible ARMA (p, q)-process $y(t) = \sum_{k=1}^{p} A_k y(t-k) + \varepsilon(t) + \sum_{k=1}^{q} B_k \varepsilon(t-k)$. First observe that $y(t) - \hat{y}(t|t-1) = \varepsilon(t)$, and that $\hat{y}(t|t-1)$ is linear in $\varepsilon(s)$ and $y(s)$, $s < t$. Note that $E(\varepsilon(t)|y(s); s < t) = 0$, $E(y(t-k)|y(s); s < t) = y(t-k)$, and $E(\hat{y}(t-k|t-k-1)|y(s); s < t) = \hat{y}(t-k|t-k-1)$. Thus

$$\hat{y}(t \mid t - 1) = \sum_{k=1}^{p} A_k y(t - k) + \sum_{k=1}^{q} B_k \{y(t - k) - \hat{y}(t - k \mid t - k - 1)\}. \qquad (7.5)$$

Note that this is a recursive expression for the one step ahead predictor. Similarly we obtain for the two-step ahead predictor

$$\hat{y}(t \mid t - 2) = A_1 \hat{y}(t - 1 \mid t - 2) + \sum_{k=2}^{p} A_k y(t - k)$$
$$+ \sum_{k=2}^{q} B_k \{y(t - k) - \hat{y}(t - k \mid t - k - 1)\}.$$

An analogous formula holds for the m-step ahead predictor.

7.2   Spectral Filtering

Wiener and Kolmogorov solved the filtering problem in terms of the spectral properties of the processes. Suppose that the joint process (y, z) has zero mean and that it has no perfectly predictable component. The process y then has an invertible MA representation, see Theorem 6.2.4. We denote this by $y(t) = \sum_{k=0}^{\infty} G_y(k)\varepsilon(t - k)$, where $\widehat{G_y}(z) = \sum_{k=0}^{\infty} G_y(k)z^{-k}$. Denote the autocovariances by

$$R_{yy}(k) := E(y(t)y^T(t - k)) \text{ and } R_{zy}(k) := E(z(t)y^T(t - k)),$$

and the spectra by $S_{yy}(e^{i\omega})$ and $S_{zy}(e^{i\omega})$ respectively, as defined in formula (6.27). According to Theorem 6.5.3 there holds

$$S_{yy}(e^{i\omega}) = \frac{1}{2\pi} \widehat{G_y}(e^{i\omega}) \widehat{G_y}^T(e^{-i\omega}).$$

We shall assume here that the information given to us regarding the processes y and z is the spectra S_{yy} and S_{zy}, or equivalently, the autocovariances R_{yy} and R_{zy}. Although, as already stated, the process y has an invertible MA representation $y(t) = \sum_{k=0}^{\infty} G_y(k)\varepsilon(t - k)$ we do not assume that the function $\widehat{G_y}(z)$ is known. In fact, it turns out that finding this function is the key step in solving the m-step ahead filtering (or smoothing) problem.

The solution of the filtering problem in terms of the spectral properties in the unrestricted smoothing case are covered by the next result.

Proposition 7.2.1 *For the joint Gaussian process (y, z) with zero mean and no predictable component and given spectra S_{yy} and S_{zy}, we define*

$$\widehat{F}(e^{i\omega}) = S_{zy}(e^{i\omega})S_{yy}^{-1}(e^{i\omega}) = \sum_{k=-\infty}^{\infty} F(k)e^{-ik\omega}.$$

Then $\hat{z}(t) = \sum_{k=-\infty}^{\infty} F(k)y(t-k)$, is the best linear predictor for $z(t)$ based on all $y(s)$.

Proof The prediction $\hat{z}(t)$ should be such that $E(z(t) - \hat{z}(t))y^T(s) = 0$ for all s. So for each k the condition is that

$$R_{zy}(k) = E\hat{z}(t)y^T(t-k) = E\left(\sum_{j=-\infty}^{\infty} F(j)y(t-j)\right)y^T(t-k)$$

$$= \sum_{j=-\infty}^{\infty} F(j)Ey(t-j)y^T(t-k) = \sum_{j=-\infty}^{\infty} F(j)R_{yy}(k-j) \qquad k \in \mathbb{Z}.$$

The very left hand side of this equality is the coefficient of $e^{ik\omega}$ in $S_{zy}(e^{i\omega})$ and the right hand side is the coefficient of $e^{-ik\omega}$ in $\widehat{F}(e^{i\omega})S_{yy}(e^{i\omega})$. □

The set of equations $R_{zy}(k) = \sum_{j=-\infty}^{\infty} F(i)R_{yy}(k-j)$ for the unknown $F(j)$ is sometimes called the *discrete time Wiener-Hopf equation*.

The solution of the m-step ahead prediction problem needs a factorization of the spectrum of the observed process y. That is, we need the function $\widehat{G}_y$ explicitly. The problem to recover the MA filter $\widehat{G}_y$ from the spectrum $S_{yy}(e^{i\omega}) = \frac{1}{2\pi}\widehat{G}_y(e^{i\omega})\widehat{G}_y^T(e^{-i\omega})$ is called the *spectral factorization problem*. In particular, we wish to determine the so-called Wold factor, that is $\widehat{G}_y$ should be causal and should have a causal inverse. For the case where S_{yy} is a rational matrix valued function this means that G_y should be a rational matrix valued function with all its poles in the open unit disc, and its inverse should also have all its poles in the open unit disc. The problem can be solved by means of state space techniques for rational spectra, as we will see in the next section.

Before stating the theorem, we introduce the following notation: for a formal power series $H(z) := \sum_{k=-\infty}^{\infty} H_k z^{-k}$ we use the notation $[H]_m^+(z) := \sum_{k=m}^{\infty} H_k z^{-k}$.

Theorem 7.2.2 (i) *For the joint Gaussian process (y, z) with zero mean and no predictable component and given spectra S_{yy} and S_{zy}, assume that G_y is a causal filter with causal inverse such that $S_{yy}(e^{i\omega}) = \frac{1}{2\pi}\widehat{G}_y(e^{i\omega})\widehat{G}_y^T(e^{-i\omega})$. We define*

$$\widehat{F_m}(e^{i\omega}) = \sum_{k=m}^{\infty} F_m(k)e^{-ik\omega} = [2\pi S_{zy}(e^{i\omega})\{\widehat{G}_y^T(e^{i\omega})\}^{-1}]_m^+\{\widehat{G}_y(e^{i\omega})\}^{-1}.$$

Then $\hat{z}(t) = \sum_{k=m}^{\infty} F_m(k)y(t-k)$, is the optimal m-step ahead predictor.
(ii) *If $z(t) = y(t)$, then*

$$\widehat{F_m}(e^{i\omega}) = \sum_{k=m}^{\infty} F_m(k)e^{-ik\omega} = [\widehat{G_y}^T(e^{i\omega})]_m^+\{\widehat{G_y}(e^{i\omega})\}^{-1}$$

gives the optimal m-step ahead predictor, that is

$$\hat{y}(t \mid t-m) = \sum_{k=m}^{\infty} G_y(k)\varepsilon(t-k).$$

Proof   The m-step ahead prediction $\hat{z}(t)$ should be orthogonal to the space of available observations, that is, $E\{z(t) - \hat{z}(t)\}y^T(s) = 0$ for all $s \leq t-m$. So for $k \geq m$ the condition is that $R_{zy}(k) = E\hat{z}(t)y^T(t-k) = 0$. Let $\widehat{F_\infty}(e^{i\omega}) = S_{zy}(e^{i\omega})S_{yy}^{-1}(e^{i\omega})$ According to Proposition 7.2.1 the process $F_\infty y$ is the optimal linear prediction of z based on y. Hence, $\eta = z - F_\infty y$ is uncorrelated with y. Put $y = G_y\varepsilon$ with ε a standard white noise and $\hat{H} = \widehat{F_\infty G_y}$. Then $\eta = z - H\varepsilon$ and η is uncorrelated with ε. Defining $G_y(k) = 0$ for $k < 0$ and using $\widehat{G_y}(e^{i\omega}) = \sum_{k=0}^{\infty} G_y(k)(e^{-ik\omega})$, we get that

$$E\|z(t) - \sum_{k=m}^{\infty} F_m(k)y(t-k)\|^2$$

$$= E\|\eta(t)\|^2 + E\|\sum_{j=-\infty}^{\infty} H(j)\varepsilon(t-j) - \sum_{k=m}^{\infty}\sum_{j=0}^{\infty} F_m(k)G_y(j)\varepsilon(t-k-j)\|^2$$

$$= E\|\eta(t)\|^2 + E\|\sum_{j=-\infty}^{\infty} H(j)\varepsilon(t-j) - \sum_{j=m}^{\infty}\{\sum_{k=m}^{\infty} F_m(k)G_y(j-k)\}\varepsilon(t-j)\|^2.$$

As the process ε is white noise, the optimal solution is obtained by taking

$$\sum_{k=m}^{\infty} F_m(k)G_y(j-k) = H(j) \text{ for all } j \geq m.$$

This means that the coefficients in $\widehat{F_m}(e^{i\omega})\widehat{G_y}(e^{i\omega})$ should coincide with those in $\hat{H}(e^{i\omega})$ for all terms $e^{-ij\omega}$ with $j \geq m$, i.e., $\widehat{F_m}(e^{i\omega})\widehat{G_y}(e^{i\omega}) = [\hat{H}(e^{i\omega})]_m^+$.

Finally, to prove (ii), apply result (i) with $z = y$. Notice that we get $\hat{H} = \widehat{G_y}$ and hence $\widehat{F_m}(e^{i\omega})\widehat{G_y}(e^{i\omega}) = [\widehat{G_y}(e^{i\omega})]_m^+$. Hence $\hat{y} = Fy = FG_y\varepsilon = [G_y]_m^+\varepsilon$, which means $\hat{y}(t \mid t-m) = \sum_{k=m}^{\infty} G_y(k)\varepsilon(t-k)$. This is in agreement with (7.4). $\square$

In the case of univariate rational spectra, i.e., univariate ARMA processes, the spectral factor can be constructed from formula (6.33). We illustrate this by a simple example.

Example 7.2.1 Consider the ARMA(1,1) model $y(t) = ay(t-1) + \varepsilon(t) + b\varepsilon(t-1)$ with $\sigma_\varepsilon^2 = 1$. Assume that this model is causal and invertible, so that, $-1 < a < 1$ and $-1 < b < 1$. In this case the spectral factor is $\hat{G}(e^{i\omega}) = \frac{b(e^{i\omega})}{a(e^{i\omega})} = \{1 + be^{-i\omega}\}/\{1 - ae^{-i\omega}\} = 1 + (a+b)\sum_{k=1}^{\infty} a^{k-1}e^{-ik\omega}$. Here a and b are easily determined from the spectrum in (6.33), as a is the stable pole and b is the stable zero of this function.

So for $m \geq 1$ the m-step ahead predictor is given by

$$[\hat{G}(e^{i\omega})]_m^+ \hat{G}^{-1}(e^{i\omega}) = (a+b) \sum_{k=m}^{\infty} a^{k-1} e^{-ik\omega} \cdot \frac{1 - ae^{-i\omega}}{1 + be^{-i\omega}} =$$

$$= (a+b)a^{m-1}e^{-im\omega} \sum_{k=0}^{\infty} (-b)^k e^{-ik\omega}.$$

So the optimal predictor of $y(t+m)$ based on the data $\{y(s); s \leq t\}$ is obtained from

$$\hat{y}(t+m \mid t) = (a+b)a^{m-1} \sum_{k=0}^{\infty} (-b)^k y(t-k).$$

The variance of the prediction error is (see Proposition 7.1.2)

$$1 + \sum_{k=1}^{m-1} \{(a+b)a^{k-1}\}^2 = 1 + (a+b)^2 \cdot \frac{1 - a^{2m-2}}{1 - a^2}$$

For $m \to \infty$ this tends to the unconditional variance $1 + (a+b)^2/(1-a^2)$ of the process, as would be expected.

Note that for an MA(1) process, with $a = 0$, one has

$$\hat{y}(t+1 \mid t) = b \sum_{k=0}^{\infty} (-b)^k y(t-k) = b\varepsilon(t)$$

and $\hat{y}(t+m \mid t) = 0$ for $m \geq 2$, which reflects that process values more than one time unit apart are uncorrelated. For an AR(1) process, with $b = 0$, it follows that $\hat{y}(t+m \mid t) = a^m y(t)$. These results are also easily obtained from Proposition 7.1.2.

7.3   The Kalman Filter

An efficient approach to filtering and prediction was developed by Kalman and Bucy. Here the starting point is not a spectral representation of the process, but a state space model, that is

$$\begin{cases} x(t+1) & = Ax(t) + F\varepsilon(t) \\ y(t) & = C_1 x(t) + G_1\varepsilon(t) \\ z(t) & = C_2 x(t) + G_2\varepsilon(t) \end{cases} \tag{7.6}$$

We make the following assumptions. The white noise process ε has zero mean and covariance $E\varepsilon(t)\varepsilon^T(t) = I$, which can always be achieved by appropriate transformations. The matrices in (7.6) are given. However, we do not require that A is a stable matrix, that is, the processes need not be causal. The process y is assumed to have no perfectly predictable component. For this reason we assume that G_1 has full row rank, i.e., has linearly independent rows. We further assume that observations $y(t)$ are available for $t \geq 0$, and that the initial condition $x(0)$ is a zero mean Gaussian random variable with covariance matrix $P(0)$. Finally, we assume that $x(0)$ is independent of $\varepsilon(t)$ for $t \geq 0$.

As before, we consider the problem (7.1) of optimal filtering of z on the basis of observations from y. For simplicity we restrict the attention to Gaussian processes, because the solution for this case is also optimal among the linear predictors for arbitrary distributions. As $E\varepsilon(t)y^T(s) = 0$ for all $s < t$ it follows from (7.6) that the optimal filter $\hat{z}(t)$ is given by $\hat{z}(t) = C_2\hat{x}(t)$, where

$$\hat{x}(t) = E(x(t) \mid y(s), 0 \leq s \leq t - 1) \tag{7.7}$$

So the filtering problem can be expressed in terms of the question how to predict the state in (7.6) from the observations of the process y. As this does not cause any additional problems, we will consider this *state filtering problem* for ARMAX systems with exogenous inputs, that is, for which the inputs are completely uncorrelated with the outputs

$$Eu(t)y^T(s) = 0 \quad \text{for all} \quad t, s \geq 0$$

So we consider systems of the form

$$x(t+1) = Ax(t) + Bu(t) + F\varepsilon(t) \tag{7.8}$$

$$y(t) = Cx(t) + Du(t) + G\varepsilon(t) \tag{7.9}$$

where the white noise process ε has zero mean and unit covariance matrix, and where the matrix G has full row rank. We remark that if x in (7.6) is a minimal state for the joint process (y, z), then it is also a state for the process y, but in general not a minimal one. Also the number of auxiliary noise variables ε is in general not minimal for the representation of y. We will show that the process $\hat{x}$ of (7.7) is also a (non-minimal) state for y, although in general not for (y, z).

The solution (7.7) of the filtering problem for the system (7.8), (7.9) is given by the *Kalman-Bucy filter*, also referred to as *Kalman filter*. We use the notation

$$\hat{y}(t) := E(y(t) \mid y(s); 0 \le s \le t - 1) = C\hat{x}(t) + Du(t) \tag{7.10}$$

and denote the corresponding prediction error by

$$\omega(t) = y(t) - \hat{y}(t) = y(t) - C\hat{x}(t) - Du(t). \tag{7.11}$$

This is the forward innovations process. Further we denote the covariance matrix of the state reconstruction error by

$$P(t) = E\big(x(t) - \hat{x}(t)\big)\big(x(t) - \hat{x}(t)\big)^T$$

The following result gives recursive formulas for the computation of $\hat{x}$ and P, which are of immediate use in prediction and filtering.

Theorem 7.3.1 *The optimal filter for the state is given by*

$$\hat{x}(t + 1) = A\hat{x}(t) + Bu(t) + K(t)\omega(t), \quad \hat{x}(0) = 0, \tag{7.12}$$

where

$$\omega(t) = y(t) - C\hat{x}(t) - Du(t),$$

and where $K(t)$ is defined recursively in terms of $P(t)$ as follows

$$K(t) = \{AP(t)C^T + FG^T\}\{CP(t)C^T + GG^T\}^{-1}, \tag{7.13}$$

$$P(t + 1) = \{A - K(t)C\}P(t)\{A - K(t)C\}^T + \{F - K(t)G\}\{F - K(t)G\}^T, \tag{7.14}$$

where $P(0)$ is the covariance matrix of $x(0)$.

Before proving the theorem, let us comment on the recursive nature of the filter computation. First of all, given $P(0)$, the matrices $P(t)$ and $K(t)$ can be computed independent of observations of the process $y(t)$ and the input $u(t)$. Then, as soon as we know $u(0)$ and have observed $y(0)$ we also know $\omega(0) = y(0) - Du(0)$. From this we can compute $\hat{x}(1)$. Once we have the input $u(1)$ we can also compute $\hat{y}(1)$. Subsequently, after observing $y(1)$ we can compute $\omega(1) = y(1) - C\hat{x}(1) - Du(1)$, as well as $\hat{x}(2)$, and we continue in this fashion.

Note that the optimal filter has the structure of a state observer.

Let us also comment on the choice of $P(0)$, which may or may not be given. In case $P(0) = 0$ we have $x(0) = 0$ and conversely, if $x(0)$ is deterministic, we have $P(0) = 0$. In case $P(0)$ is unknown one can make appropriate choices, which we shall comment upon in the next section.

Proof As the inputs are exogenous they can be considered as fixed. For simplicity we assume that $u(t) = 0$ for all $t \geq 0$ and that $P(0)$ is given as the covariance matrix of $x(0)$. We put $\hat{x}(0) = 0$, motivated by $\hat{x}(0) = E(x(0) \mid y(s), 0 \leq s \leq -1) = 0$. The proof for the general case can be done in an analogous way. Under these assumptions, $\varepsilon, x, y, \hat{x}$ and ω all are Gaussian process with zero mean. It follows from (7.8), (7.9) and (7.11) that the components of $\{y(s), s \leq t\}$, and those of $\{\omega(s), s \leq t\}$ span the same subspace of random variables as the components of $\{\varepsilon(s), s \leq t\}$. Remark that $E\omega(t)\omega(s)^T = 0$ for all $s \neq t$.

To prove (7.12) we make the following computation

$$
\begin{aligned}
\hat{x}(t+1) &= E\big(x(t+1) \mid y(s), s \leq t\big) = E\big(x(t+1) \mid \omega(s), s \leq t\big) \\
&= E\big(x(t+1) \mid \omega(s), s \leq t-1\big) + E\big(x(t+1) \mid \omega(t)\big) \\
&= E\big(Ax(t) + F\varepsilon(t) \mid \omega(s), s \leq t-1\big) + E\big(x(t+1) \mid \omega(t)\big) \\
&= A\hat{x}(t) + E\big(x(t+1) \mid \omega(t)\big).
\end{aligned}
$$

Here we used in the second line the orthogonality of $\omega(t)$ and $\{\omega(s), s \leq t-1\}$, and in the third line the fact that $\varepsilon(t)$ is uncorrelated with $\{\varepsilon(s), s \leq t-1\}$, hence also with $\{\omega(s), s \leq t-1\}$. According to the second paragraph of Sect. 6.4 we have that

$$
E\big(x(t+1) \mid \omega(t)\big) = E\big(x(t+1)\omega(t)^T\big)E\big(\omega(t)\omega(t)^T\big)^{-1}\omega(t).
$$

So to prove (7.12) and (7.13) it is sufficient to show that

$$
E\omega(t)\omega(t)^T = CP(t)C^T + GG^T, \tag{7.15}
$$

$$
Ex(t+1)\omega(t)^T = AP(t)C^T + FG^T. \tag{7.16}
$$

To check the first formula remark that $\omega(t) = y(t) - \hat{y}(t) = C(x(t) - \hat{x}(t)) + G\varepsilon(t)$, and use that $x(t) - \hat{x}(t)$ and $\varepsilon(t)$ are uncorrelated because $x(t)$ and $\hat{x}(t)$ are linear functions of

$\{\varepsilon(s), s \leq t - 1\}$. For the second formula we observe

$$
\begin{aligned}
Ex(t+1)\omega(t)^T &= E\big(Ax(t) + F\varepsilon(t)\big)\big(Cx(t) + G\varepsilon(t) - C\hat{x}(t)\big)^T \\
&= E\big(A(x(t) - \hat{x}(t)) + A\hat{x}(t) + F\varepsilon(t)\big)\big(C(x(t) - \hat{x}(t)) + G\varepsilon(t)\big)^T \\
&= AP(t)C^T + FG^T,
\end{aligned}
$$

because $x(t) - \hat{x}(t)$ and $\hat{x}(t)$ are uncorrelated with $\varepsilon(t)$, and because $x(t) - \hat{x}(t)$ is orthogonal to $\hat{x}(t)$. We proved (7.12) and (7.13).

The result in (7.14) follows from

$$
\begin{aligned}
x(t+1) - \hat{x}(t+1) &= A(x(t) - \hat{x}(t)) + F\varepsilon(t) - K(t)\omega(t) \\
&= A(x(t) - \hat{x}(t)) + F\varepsilon(t) - K(t)(C(x(t) - \hat{x}(t)) + G\varepsilon(t)) \\
&= (A - K(t)C)(x(t) - \hat{x}(t)) + (F - K(t)G)\varepsilon(t)
\end{aligned}
$$

and the fact that $x(t) - \hat{x}(t)$ and $\varepsilon(t)$ are uncorrelated. $\quad\square$

Rewrite Eqs. (7.12) and (7.11) as

$$
\hat{x}(t+1) = A\hat{x}(t) + Bu(t) + K(t)\omega(t) \tag{7.17}
$$

$$
y(t) = C\hat{x}(t) + Du(t) + \omega(t) \tag{7.18}
$$

This is a state space model for the process y, with state $\hat{x}$ and with the innovations ω as driving noise process. The state updating Eq. (7.17) expresses the new state in terms of the predicted part, $A\hat{x}(t) + Bu(t)$, and an adjustment based on the prediction error $\omega(t)$. The so-called *Kalman gain* $K(t)$ measures the extent in which this new information is taken into account. The filter is recursive and the matrix recursions (7.13), (7.14) are independent of the data. This means that the Kalman gain $K(t)$ and the error covariances $P(t)$ can be computed off-linc, before the actual observations are coming in. This is an attractive property for applications that require fast updating.

The Kalman filter can be applied directly in prediction.

Proposition 7.3.2 *The optimal one-step ahead predictor is given by*

$$
\hat{y}(t) = C\hat{x}(t) + Du(t),
$$

with covariance matrix of the prediction error equal to

$$
E\left(\big((y(t) - \hat{y}(t))(y(t) - \hat{y}(t))^T\big)\right) = CP(t)C^T + GG^T.
$$

The optimal m-step ahead predictor (for given inputs) is

$$\hat{y}(t+m-1 \mid t-1) = CA^{m-1}\hat{x}(t) + Du(t+m-1) + \sum_{j=1}^{m-1} CA^{j-1}Bu(t+m-1-j)$$

with prediction error covariance

$$CA^{m-1}P(t)(A^T)^{m-1}C^T + GG^T + \sum_{j=1}^{m-1} CA^{j-1}GG^T(A^T)^{j-1}C^T.$$

Proof The results on the one step ahead predictor appeared already in the formulas (7.10) and (7.15).

For m-step ahead prediction note that (7.8) and (7.9) imply that

$$\begin{aligned} y(t+m-1) &= CA^{m-1}x(t) + Du(t+m-1) + G\varepsilon(t+m-1) \\ &\quad + \textstyle\sum_{j=1}^{m-1} CA^{j-1}\big(Bu(t+m-1-j) + G\varepsilon(t+m-1-j)\big). \end{aligned}$$

Since $\varepsilon(j)$ is uncorrelated with $y(s)$ for $j \geq s$ we have

$$\begin{aligned} \hat{y}(t+m-1 \mid t-1) &= CA^{m-1}E\big(x(t) \mid y(s), s < t\big) + Du(t+m-1) \\ &\quad + \textstyle\sum_{j=1}^{m-1} CA^{j-1}\big(Bu(t+m-1-j)\big). \end{aligned}$$

This proves the formula for $\hat{y}(t+m-1 \mid t-1)$. Furthermore

$$\begin{aligned} y(t+m-1) - \hat{y}(t+m-1 \mid t-1) &= CA^{m-1}(x(t) - \hat{x}(t)) + G\varepsilon(t+m-1) \\ &\quad + \textstyle\sum_{j=1}^{m-1} CA^{j-1}G\varepsilon(t+m-1-j). \end{aligned}$$

Again use that $x(t) - \hat{x}(t)$ is uncorrelated with $\varepsilon(s)$ for $s \geq t$ to obtain the formula for covariance of $y(t+m-1) - \hat{y}(t+m-1 \mid t-1)$, i.e., the prediction error covariance. $\square$

The Kalman filter can also be used in smoothing and filtering. Let observations $\{y(t), t = 0, \dots, N\}$ be available and suppose we wish to determine the smoothed value $\hat{x}(t_0 \mid N) = E\big(x(t_0) \mid y(t), t = 0, \dots, N\big)$, for some $0 \leq t_0 \leq N$. Define the extended state by $x_e(t) = \big(x(t)^T \; x(t_0)^T\big)^T$, then we can rewrite (7.8) and (7.9) in terms of this extended state with parameters

$$A_e = \begin{pmatrix} A & 0 \\ 0 & I \end{pmatrix}, \quad B_e = \begin{pmatrix} B \\ 0 \end{pmatrix}, \quad C_e = \begin{pmatrix} C & 0 \end{pmatrix}, \quad D_e = D, \quad F_e = \begin{pmatrix} F \\ 0 \end{pmatrix}, \quad G_e = G.$$

Applying the Kalman filter to this extended system, we obtain $\hat{x}_e(N+1) = E(x_e(N+1) \mid y(t), t = 0, \ldots, N)$, and therefore also $\hat{x}(t_0 \mid N)$. In fact, it is not necessary to run the Kalman filter separately for every time instant t_0. All smoothed values $\hat{x}(t \mid N)$ can be calculated by first applying the Kalman filter, followed by a backward recursion starting from the final filtered state $\hat{x}(N+1)$ in (7.12). For algorithmic details we refer to [24], [1].

The following result solves the true filtering problem, that is, the best state estimate based on the past and current observations.

Proposition 7.3.3 *The filtered state* $\hat{x}(t \mid t) := E\{x(t) \mid y(s); 0 \leq s \leq t\}$ *and its error covariance* $P(t \mid t) := E\left((x(t) - \hat{x}(t \mid t))(x(t) - \hat{x}(t \mid t))^T\right)$ *are given by*

$$\hat{x}(t \mid t) = \hat{x}(t) + P(t)C^T\{CP(t)C^T + GG^T\}^{-1}\omega(t) \tag{7.19}$$

$$P(t \mid t) = P(t) - P(t)C^T\{CP(t)C^T + GG^T\}^{-1}CP(t) \tag{7.20}$$

Proof The space spanned by the components of $\{y(s); 0 \leq s \leq t\}$ can be decomposed into the two orthogonal components, the first spanned by the components of $\{y(s); 0 \leq s \leq t - 1\}$ and the second spanned by the components of $\{\omega(t)\}$. Because of this, and using that $(x(t), \omega(t))$ has a joint Gaussian distribution, it follows that

$$\hat{x}(t \mid t) = E\left(x(t) \mid y(s), s \leq t - 1\right) + E\left(x(t) \mid \omega(t)\right) =$$
$$= \hat{x}(t) + E\left(x(t)\omega(t)^T\right)E\left(\omega(t)\omega(t)^T\right)^{-1}\omega(t).$$

Here $E\left(\omega(t)\omega(t)^T\right) = CP(t)C^T + GG^T$ and

$$Ex(t)\omega(t)^T = E\left((x(t) - \hat{x}(t))(Cx(t) - C\hat{x}(t) + G\varepsilon(t))^T\right) = P(t)C^T,$$

because $E\hat{x}(t)\omega(t)^T = 0$, $Ex(t)\varepsilon^T(t) = 0$ and $E\hat{x}(t)\varepsilon^T(t) = 0$. This proves (7.19). Further, let $L(t) := P(t)C^T\{CP(t)C^T + GG^T\}^{-1}$, and recall that $E(\omega(t)\omega(t)^T) = CP(t)C^T$ and $E(x(t)\omega(t)^T) = P(t)C^T$. Then

$$P(t \mid t) = E\left((x(t) - \hat{x}(t) - L(t)\omega(t))(x(t) - \hat{x}(t) - L(t)\omega(t))^T\right)$$
$$= P(t) + L(t)E(\omega(t)\omega(t)^T)L(t)^T -$$
$$\quad -E\left((x(t) - \hat{x}(t))\omega^T(t)\right)L(t)^T - L(t)E\left(\omega(t)(x(t) - \hat{x}(t))^T\right)$$
$$= P(t) + P(t)C^TL(t)^T - P(t)C^TL(t)^T - L(t)CP(t)$$
$$= P(t) - L(t)CP(t),$$

where we also used the fact that $E\hat{x}(t)\omega^T(t) = 0$. This shows (7.20). $\qquad\square$

Example 7.3.1 Consider again the ARMA(1,1) model $y(t) = ay(t-1) + \varepsilon(t) + b\varepsilon(t-1)$ with $\sigma_\varepsilon^2 = 1$. Define $x(t) = ay(t-1) + b\varepsilon(t-1)$, then $y(t) = x(t) + \varepsilon(t)$ and $x(t+1) = ay(t) + b\varepsilon(t) = ax(t) + (a+b)\varepsilon(t)$. So a state space representation is obtained by defining the parameters in (7.8), (7.9) by

$$A = a, \ B = 0, \ C = 1, \ D = 0, \ F = a+b, \ G = 1.$$

Suppose that $x(0)$ is a zero mean Gaussian random variable with variance $p(0)$, then the filter equations (7.12), (7.13) and (7.14) are given by

$$\hat{x}(t+1) = a\hat{x}(t) + k(t)(y(t) - \hat{x}(t)) = (a - k(t))\hat{x}(t) + k(t)y(t),$$

$$k(t) = (ap(t) + a + b)/(p(t) + 1),$$

$$p(t+1) = p(t)(a - k(t))^2 + (a + b - k(t))^2.$$

We consider the case of an MA(1) process in more detail. Then $a = 0$ and $y(t) = \varepsilon(t) + b\varepsilon(t-1)$. By using the values of A, B, C, D, F and G given above and substituting (7.13) in (7.14) it follows that $p(t+1) = b^2 p(t)/(1 + p(t))$, or more explicitly

$$p(t+1) = \frac{p(1)b^{2t}}{(1 + p(1)\sum_{j=0}^{t-1} b^{2j})}.$$

If the MA process is invertible, that is, if $|b| < 1$, then $p(t) \to 0$ as $t \to \infty$. So, in the limit, we can reconstruct the state $x(t) = b\varepsilon(t-1)$ without error from the information $\{y(s), 0 \leq s \leq t-1\}$. We also refer to Example 6.3.1, in this case $\varepsilon(t) = \sum_{k=0}^\infty (-b)^k y(t-k)$, so that $\varepsilon(t-1)$ is a function of $\{y(s), s \leq t-1\}$. The error occurs because the observations $\{y(s), s \leq -1\}$ are not available, but this error disappears in the limit when $|b| < 1$.

If $b = \pm 1$, then still $p(t) \to 0$ for $t \to \infty$. On the other hand, if $|b| > 1$, then by rewriting $p(t+1) = p(1)b^{2t}(1-b^2)/(1-b^2+p(1)(1-b^{2t}))$ it follows that $p(t) \to b^2 - 1$ for $t \to \infty$. So in this case the error does not vanish in the limit.

Example 7.3.2 Suppose that z is a random process that is observed under noise. We assume that

$$z(t+1) = z(t) + \varepsilon_1(t), \ \ y(t) = z(t) + \varepsilon_2(t),$$

where $(\varepsilon_1(t), \varepsilon_2(t))^T$ is a bivariate Gaussian white noise process with mean zero and covariance matrix $\begin{pmatrix} \sigma_1^2 & 0 \\ 0 & \sigma_2^2 \end{pmatrix}$. Here y is observed, but z is unobserved. This was proposed, for instance as a possible model of price formation in Example 1.1.1, where y denotes the observed price and z the underlying fundamental price that is affected by random variations in the market. The aim here is to construct an optimal estimate of the fundamental price on the basis of past observations, that is, the filtering problem $\hat{z}(t) = E(z(t) \mid y(s), 0 \le s \le t - 1)$. This is solved by the Kalman filter, where the matrices in (7.8) and (7.9) are given by

$$A = 1, \quad B = 0, \quad C = 1, \quad D = 0, \quad F = \begin{pmatrix} \sigma_1 & 0 \end{pmatrix}, \quad G = \begin{pmatrix} 0 & \sigma_2 \end{pmatrix}.$$

Using these values and substituting (7.13) in (7.14) one finds that the filter formulas are

$$k(t) = \frac{p(t)}{p(t) + \sigma_2^2}, \qquad p(t+1) = \frac{p(t)(\sigma_1^2 + \sigma_2^2) + \sigma_1^2 \sigma_2^2}{p(t) + \sigma_2^2},$$

and the optimal estimate is given by

$$\hat{z}(t+1) = \hat{z}(t) + k(t)(y(t) - \hat{z}(t)) = (1 - k(t))\hat{z}(t) + k(t)y(t).$$

This is also called an adaptive expectations model, where the expectations $\hat{z}(t)$ are updated because of the prediction errors $y(t) - \hat{z}(t)$. If $t \to \infty$ then for every $p(0) > 0$ the sequence $p(t)$ converges to the positive solutions of $p = ((\sigma_1^2 + \sigma_2^2)p + \sigma_1^2 \sigma_2^2)/(p + \sigma_2^2)$, that is, to $p = \frac{1}{2}(\sigma_1^2 + \sqrt{\sigma_1^4 + 4\sigma_1^2 \sigma_2^2})$. The corresponding gain is

$$k = \frac{\sigma_1^2 + \sqrt{\sigma_1^4 + 4\sigma_1^2 \sigma_2^2}}{\sigma_1^2 + 2\sigma_2^2 + \sqrt{\sigma_1^4 + 4\sigma_1^2 \sigma_2^2}}.$$

So $0 < k < 1$, and k is small if σ_2^2 is large relative to σ_1^2, and k is large if σ_2^2 is small relative to σ_1^2. In the limit, the adaptive expectations model can be rewritten as the following forecast model for the process y

$$\hat{y}(t+1) = \hat{z}(t+1) = \hat{z}(t) + k(y(t) - \hat{z}(t)) = k \sum_{j=0}^{\infty} (1-k)^j y(t-j).$$

This is also called the method of exponentially weighted moving averages for forecasting. The forecast series $\hat{y}(t)$ is smooth if $k \approx 0$, that is, if the variance σ_2^2 in the observations is large relative to the variance σ_1^2 in the underlying process. On the other hand, if $k \approx 1$, so that the observation variance σ_2^2 is relatively small, then $\hat{y}(t) \approx y(t-1)$ and so the up- and down- movements of the observed series are followed fast.

7.4   The Steady State Filter

The application of the Kalman filter requires initial values $\hat{x}(0)$ and $P(0)$ in the recursion (7.12) and (7.14). If the observed series is relatively short then the results may be sensitive with respect to these initial values. Their specification becomes less important if the number of observations increases. Under appropriate conditions the filter becomes time-invariant and independent of the initial conditions if the number of observations tends to infinity. To make this more precise, we first state an auxiliary result.

Proposition 7.4.1 *The recursions* (7.13) *and* (7.14) *are equivalent to*

$$P(t+1) = AP(t)A^T + FF^T -$$
$$(GF^T + CP(t)A^T)^T (CP(t)C^T + GG^T)^{-1}(GF^T + CP(t)A^T) \tag{7.21}$$

Proof   The result follows by substituting (7.13) into (7.14) and rewriting the resulting expression.   □

The *filter Riccati equation* (7.21) closely resembles the control Riccati equation (5.12). The result in Theorem 5.3.2 describes the limiting properties for $t \to \infty$ of this equation. As before we assume that y has no perfectly predictable component, so that G has full row rank and GG^T is invertible.

Theorem 7.4.2 *Assume that the pair* $(A - FG^T(GG^T)^{-1}C, F - FG^T(GG^T)^{-1}G)$ *is stabilizable, and that the pair* (A, C) *is detectable. Then the following holds true.*

(i) *For any positive definite* $P(0)$*, the solution of* (7.21) *converges as* $t \to \infty$ *to a positive semidefinite matrix* P*, which does not depend on the choice of* $P(0)$*. The corresponding solution* K *of* (7.13) *is such that* $A - KC$ *is stable.*

(ii) *P is the largest Hermitian solution of the algebraic Riccati equation*

$$P = APA^T + FF^T - (GF^T + CPA^T)^T (GG^T + CPC^T)^{-1}(GF^T + CPA^T) \tag{7.22}$$

Moreover, P is positive semidefinite.

Proof   This follows directly from Theorem 5.3.2.   □

Recall that in Example 7.3.1 for $b > 1$ there are two semi definite solutions: $p = 0$ if $P(0) = 0$ and $p = b^2 - 1$ if $p(0) \neq 0$. Both $p = 0$ and $p = b^2 - 1$ are positive semidefinite solutions of the equation (7.22).

The assumptions in the theorem are satisfied, for example, if A is a stable matrix with (A, C) observable and with $FG^T = 0$. This is the case if y is a stationary process with representation (6.19), (6.20) with $\Sigma_{12} = 0$. However, the assumptions in the theorem are far more general.

Note that the theorem also gives insight in the choice of $P(0)$ for the Kalman filter when $P(0)$ is unknown. It turns out that in case the assumptions of the theorem are satisfied the choice of $P(0)$ is immaterial to the asymptotic behaviour of the Kalman filter, provided we take $P(0)$ to be positive definite.

Under the above conditions, the Kalman filter converges to the so-called steady state filter

$$
\begin{aligned}
\hat{x}(t+1) &= A\hat{x}(t) + Bu(t) + K\omega(t), \\
y(t) &= C\hat{x}(t) + Du(t) + \omega(t),
\end{aligned}
\tag{7.23}
$$

$$
K = (APC^T + FG^T)(GG^T + CPC^T)^{-1}.
\tag{7.24}
$$

The state equation can be written as $\hat{x}(t+1) = (A - KC)\hat{x}(t) + (B - KD)u(t) + Ky(t)$, and as $A - KC$ is stable this means that the filtered state $\hat{x}$ is indeed a strictly causal function of the observed process y.

Next we consider this model without control inputs, that is, with $u(t) = 0$ for all $t \subset \mathbb{Z}$. We further assume that y is a stationary process with no perfectly predictable component and with rational spectrum. In this case the steady state filter has several interesting interpretations. It corresponds to the Wold decomposition of the process (Theorem 6.2.4), it provides a causal and invertible ARMA representation (Sect. 6.3), it solves the rational spectral factorization problem (Sect. 7.2), and it gives a state representation with minimal state covariance matrix (Theorem 6.4.5).

First we consider the Wold decomposition. So we wish to determine a moving average representation $y(t) = \sum_{k=0}^{\infty} G(k)\varepsilon(t-k)$ with causal inverse $\varepsilon(t) = \sum_{k=0}^{\infty} H(k)y(t-k)$. This can also be formulated as the spectral factorization problem, that is, the rational spectrum S of the process y should be factorized as $S(z) = \frac{1}{2\pi}G(z)G(z^{-1})^T$, where G is a rational causal function with causal inverse. Writing $G(z) = A(z)^{-1}B(z)$ with $A(z)$ and $B(z)$ polynomial matrices, this also corresponds to a causal and invertible ARMA representation of the process.

Theorem 7.4.3 *Every purely nondeterministic stationary process with rational spectrum and with no perfectly predictable component has a causal and invertible ARMA representation. The corresponding spectral factor is given by*

$$
G(z) = (I + C(zI - A)^{-1}K)(GG^T + CPC^T)^{1/2}
\tag{7.25}
$$

where (A, C, F, G) define a realization (7.8), (7.9) (with $B = 0$ and $D = 0$) and with P and K as defined in (7.22) and (7.24). The Wold decomposition of the process has filter (7.25).

Proof Because the spectrum of y is rational, it has a state space realization (7.8), (7.9) with A stable, $B = 0$ and $D = 0$. Take a minimal realization, so that in particular, (A, C) is observable, and let n be the dimension of the state space, so that A is an $n \times n$ matrix. The corresponding steady state Kalman filter realization is given by (7.23) (again, with $B = 0$, $D = 0$). Put $Q = \left(C^T \ A^T C^T \ \dots (A^T)^{n-1} C^T \right)^T$, then this matrix has full column rank n. Now

$$
\begin{pmatrix} y(t) \\ \vdots \\ y(t+n-1) \end{pmatrix} = Q\hat{x}(t) + R \begin{pmatrix} \omega(t) \\ \vdots \\ \omega(t+n-1) \end{pmatrix},
$$

where R is a matrix that can be determined explicitly in terms of A, C and K. It follows that $\hat{x}(t)$ can be expressed in terms of $y(s)$ and $\omega(s)$ for $t \leq s \leq t+n-1$. Substituting this in the state space equation we obtain an ARMA representation of the process y with ω as the driving noise process. This model is causal, as $y(t) = \omega(t) + \sum_{k=1}^{\infty} CA^{k-1}K\omega(t-k)$, and invertible, as $\omega(t) = y(t) - C\hat{x}(t)$ is by construction a function of $\{y(s), s \leq t\}$ (see (7.11)). As y is a causal function of ω and ω is also a causal function of y, this model corresponds to the Wold decomposition.

Finally, the spectral factor follows from the filter $I + C(zI - A)^{-1}K$ that produces y from ω, where ω is white noise with $\omega(t) = y(t) - C\hat{x}(t) = C(x(t) - \hat{x}(t)) + G\varepsilon(t)$, so that its covariance matrix is given by $GG^T + CPC^T$ (see also Proposition 7.3.2).   □

Now consider a process y with given covariances $\{R(k), k \in \mathbb{Z}\}$ and minimal realization (A, C, M) such that $R(k) = CA^{k-1}M$ for $k \geq 1$. The corresponding stochastic realization problem is discussed in Sect. 6.4, see in particular Theorem 6.4.5. For given (A, C, M), Π_- denotes the minimal achievable covariance matrix of the state in realizations of the process y.

Theorem 7.4.4 *For given (A, C, M) with $R(k) = CA^{k-1}M$ for $k \geq 1$, consider the steady state filter (7.23), with $B = 0$ and $D = 0$, i.e.,*

$$
\begin{aligned}
\hat{x}(t+1) &= A\hat{x}(t) + K\omega(t), \\
y(t) &= C\hat{x}(t) + \omega(t),
\end{aligned}
\tag{7.26}
$$

where

$$
K = (APC^T + FG^T)(GG^T + CPC^T)^{-1}.
\tag{7.27}
$$

Denote the covariance matrix of the state $\hat{x}$ in (7.26) by Π_-. Then Π_- is minimal among all covariance matrices of all possible state space representations with the same

(A, C, M), and Π_- satisfies the algebraic Riccati equation (6.25), that is

$$\Pi = A\Pi A^T + (M - A\Pi C^T)(R(0) - C\Pi C^T)^{-1}(M - A\Pi C^T)^T.$$

Proof For fixed (A, C, M) let (7.8), (7.9) with $B = 0$ and $D = 0$, be an arbitrary realization of the process. The optimal predictor of the state of this process is given by $\hat{x}(t) = E(x(t) \mid y(s), s \leq t - 1)$. The steady state Kalman filter (7.26), with K given by (7.27) is another realization of the process. The matrices A, C and M are the same for this representation as for the original one. For the matrix M this follows because $M = Ex(t+1)y(t)^T$, see Proposition 6.4.4, and $Ex(t+1)y(t)^T = E\hat{x}(t+1)y(t)^T$ as $\hat{x}$ is the optimal predictor, so that $\hat{x}(t+1) - x(t+1)$ is uncorrelated with $\{y(s), s \leq t\}$. Because it is an optimal predictor, it follows that the covariance matrix of $\hat{x}(t)$ is not larger than that of $x(t)$, that is, $\text{cov}(x(t)) - \text{cov}(\hat{x}(t))$ is positive semidefinite. This holds true for every realization with the same matrices A, C, M, so it follows that $\text{cov}(\hat{x}(t)) = \Pi_-$, by definition.

It remains to show that Π_- satisfies the algebraic Riccati equation (6.25). The steady state Kalman filter realization is given by (7.26). Here $\hat{x}(t)$ and $\omega(t)$ are uncorrelated, as $\omega(t) = y(t) - \hat{y}(t)$ is uncorrelated with $\{y(s), s \leq t - 1\}$. This shows that $\Pi_- = A\Pi_- A^T + K\Sigma_\omega K^T$, where $\Sigma_\omega = E\omega(t)\omega(t)^T = R(0) - C\Pi_- C^T$ is the covariance matrix of $\omega(t)$. Further, $M = E\hat{x}(t+1)y(t)^T$, which by (7.26) is equal to $E(A\hat{x}(t) + K\omega(t))(C\hat{x}(t)+\omega(t))^T = A\Pi_- C^T + K\Sigma_\omega$, so that $K = (M - A\Pi_- C^T)\Sigma_\omega^{-1}$. Combining these results we get that Π_- satisfies

$$\Pi_- = A\Pi_- A^T + (M - A\Pi_- C^T)(R(0) - C\Pi_- C^T)^{-1}(M - A\Pi_- C^T)^T,$$

that is, Π_- satisfies (6.25).   □

Note that this motivates to take for $P(0)$ in the Kalman filter the choice $P(0) = \Pi_-$ when $P(0)$ happens to be unknown, even though the long term behaviour of the Kalman filter is asymptotically the same for all $P(0)$ under the appropriate conditions.

Example 7.4.1 Consider again an MA(1) process $y(t) = \varepsilon(t) + b\varepsilon(t-1)$ with $\sigma_\varepsilon^2 = 1$ and with $|b| > 1$. This is a stationary process, but the representation is not invertible. Defining $x(t) = b\varepsilon(t-1)$, we can write $y(t) = x(t) + \varepsilon(t)$ and $x(t+1) = b\varepsilon(t)$, so that in terms of (7.8), (7.9) the parameters are given by $A = 0$, $B = 0$, $C = 1$, $D = 0$, $F = b$, $G = 1$. The Riccati equation (7.22) is given by $p = b^2 - b^2(1+p)^{-1}$, and the largest semidefinite solution is $p = b^2 - 1$, and then (7.24) gives $k = b(p+1)^{-1} = b^{-1}$. The steady state filter is $\hat{x}(t+1) = b^{-1}\omega(t)$ and $y(t) = \hat{x}(t) + \omega(t)$. By eliminating the state, we obtain the model

$$y(t) = \omega(t) + b^{-1}\omega(t-1).$$

As $|b| > 1$, this is indeed a causal and invertible representation of the process. To determine the variance σ_ω^2 of the white noise process ω, note that $Ey(t)^2 = 1 + b^2 = \sigma_\omega^2(1+b^{-2})$, so that $\sigma_\omega^2 = b^2$. The causal and causally invertible spectral factor is therefore given by $(1 + b^{-1}z)$. Further, the variance of the process $\hat{x}(t)$ is equal to $b^{-2}\sigma_\omega^2 = 1$, whereas the original state variable $x(t) = b\varepsilon(t-1)$ has variance $b^2 > 1$. In terms of (6.24), there holds $R(1) = b = CM = M$ and the variance Π of any state variable should satisfy the condition that $\begin{pmatrix} \Pi & b \\ b & 1 + b^2 - \Pi \end{pmatrix}$ is positive semidefinite. This is equivalent to the condition that $1 \leq \Pi \leq b^2$. So the steady state filter has minimal variance among all realizations.

Stochastic Control

8

Stochastic optimal control problems can in principle be solved by stochastic dynamic programming. We pay special attention to the *LQG problem* where the system is linear, the cost function is quadratic, and the random variables have Gaussian distributions. The optimal controller is given by the LQG feedback law where the unobserved state is replaced by the Kalman filter estimate.

8.1 Introduction

In this chapter we discuss control of uncertain systems. The methods discussed are mostly applied in engineering. Here one often has good knowledge of the system structure and of the costs involved in performing control actions. The situation is quite different in most economic applications. In general there is much uncertainty concerning the effect of decision variables on target variables, and there may be many objectives that are not easily quantified. This does not mean that the methods described in this chapter are of no value in economic decision making. Models of optimal control may assist in organizing relevant information, by making the objectives more explicit, and by indicating possible effects of different strategies. In this way one can get a better understanding of the source and extent of the involved uncertainties. Further, for example in business applications, firms may have a relatively clear idea of their objectives and the means which are available to them to achieve their goals.

Given the extent of uncertainty in economic decision making, simple models will often be more helpful than complex ones. We describe in this chapter some of the main ideas and techniques. We pay particular attention to the case of linear stochastic systems and quadratic control criteria. This leads to a relatively simple algorithm.

© The Author(s), under exclusive license to Springer Nature Switzerland AG 2021
C. Heij et al., *Introduction to Mathematical Systems Theory*,
https://doi.org/10.1007/978-3-030-59654-5_8

There are important differences between deterministic and stochastic control. In deterministic systems, the future development is completely determined by the current state and the future inputs. This means that every control strategy leads to known costs. In stochastic systems, however, there is an additional and unknown source of dynamics due to the disturbances. This means that, as opposed to the deterministic case, there is an essential distinction between open loop and closed loop control. A closed loop strategy may lead to lower costs, as the observed system trajectories may provide information on the disturbances and the current state of the system. This also means that the control variables play a dual role, that is, they can be manipulated to obtain additional information from the system and they should also lead to low costs. These two objectives may in general be conflicting.

If the choice of the control inputs does not affect the uncertainty about the system this is called *neutrality*. We will show that this holds true for the control of linear stochastic systems with quadratic costs. In this case the optimal controller has moreover the properties of *certainty equivalence* and *separation*. A controller is called certainty equivalent if it coincides with the controller for the deterministic system, replacing all uncertain quantities by their optimal estimates. A controller has the separation property if control and estimation do not influence each other in the following sense. The required estimates of uncertain quantities depend only on the stochastic properties of the system, not on the control objectives, and the control actions do not depend on the stochastic specification of the model.

In most applications it is difficult or even impossible to compute optimal controllers. Further, there often exists considerable uncertainty about the correct model specification and the control objectives. It may then be better to use relatively simple controllers, instead of complicated methods that are more sensitive to misspecifications. Suboptimal controllers can be based on heuristic principles, for instance separation and certainty equivalence. Such controllers are relatively easy to compute, and they may lead to acceptable performance.

8.2   Stochastic Dynamic Programming

The method of stochastic dynamic programming closely resembles the deterministic algorithm. It is again based on Bellman's principle of optimality. This requires that the control problem has finite horizon and that all the parameters of the problem are known, that is, the parameters of the cost function and the ones describing the dynamical and stochastic properties of the system.

Consider a system with control variables u, observed state vector x and unobserved disturbances ε related by

$$x(t+1) = f(t, x(t), u(t), \varepsilon(t)), \quad t = 0, \ldots, N-1 \tag{8.1}$$

Here the system function f, the horizon N, the initial state $x(0)$ and the probability distribution of ε are all assumed to be known. We impose the condition that

$$p\big(\varepsilon(t)|x(s), u(s), \varepsilon(s-1), s \leq t\big) = p\big(\varepsilon(t)|x(t), u(t)\big). \tag{8.2}$$

The input at time t may depend on the available information $\{x(s); s \leq t\}$, and it may further be restricted to belong to a set $u(t) \in U\{t, x(t)\}$. The costs in period t are given by $g_t = g(t, x(t), u(t), \varepsilon(t))$, for $t = 0, \ldots, N-1$, while for $t = N$ the cost only depends on $x(N)$. As the costs are random variables there does not exist, in general, a control policy that minimizes the costs for all possible disturbances. As criterion one often takes the expected cost. In some cases the variance of the outcomes may also be of importance, and this can easily be incorporated in the cost function by appropriate definition of the functions g_t. The cost of a control policy $u = \{u(t), t = 0, \ldots, N-1\}$ is given by

$$J(u) = E\big(g(N, x(N)) + \sum_{t=0}^{N-1} g(t, x(t), u(t), \varepsilon(t))\big). \tag{8.3}$$

Here certain (measurability) conditions on the control law have to be imposed in order that the expectation in (8.3) is well-defined. For the rather simple cases that we will consider in this chapter this causes no problems, but in more complicated problems it may. As in the deterministic case the objective is to find an input sequence $u^*(0), \ldots, u^*(N-1)$ that minimizes $J(u)$ (if possible).

The dynamic programming algorithm can be expressed, as in the deterministic case, in terms of the optimal-cost-to-go functions

$$J_N(x(N)) = g(N, x(N)), \tag{8.4}$$

$$J_t(x(t)) = \inf_{u(t)} E\big(g_t + J_{t+1}(x(t+1)) \mid x(t), u(t)\big), \quad t = N-1, \ldots, 0. \tag{8.5}$$

The optimality principle states that if $u^* = \{u^*(t), t = 0, \ldots, N-1\}$ is an optimal control policy, then the truncated policy $\{u^*(t), t = t_0, \ldots, N-1\}$ is also optimal for the system starting at $x(t_0)$ at time t_0 and with horizon N. This is expressed in the following theorem, which is completely analogous to the deterministic result in Theorem 5.2.1.

Theorem 8.2.1 *Let $u^*(t)$ achieve the infimum in* (8.5), *then the optimal control law for* (8.3) *is given by $u^* = \{u^*(t), t = 0, \ldots, N-1\}$ and the minimal cost is $J_0(x(0))$.*

Proof Let E_t denote the conditional expectation with respect to $\{x(t), u(t)\}$. For fixed initial state we obtain, by repeated conditioning and using (8.2), that $J(u) = E_0\big(g_0 + E_1\big(g_1 + \ldots + E_{N-1}(g_{N-1} + g_N) \ldots\big)\big)$. Because the current input only affects the current and future costs this implies that

$$\inf_u J(u) = \inf_{u(0)} E_0\big(g_0 + \inf_{u(1)} E_1\big(g_1 + \ldots + \inf_{u(N-1)} E_{N-1}(g_{N-1} + g_N)\ldots\big)\big),$$

so the minimal cost is obtained by solving (8.5) for $t = N - 1, \ldots, 0$ with cost $J_0(x(0))$.

$\square$

Although the optimality principle seems quite trivial it should be mentioned that it states sufficient conditions for optimality, not necessary ones.

The stochastic dynamic programming algorithm solves (8.3) by iteratively solving the simpler, non-dynamic optimization problems (8.5), $t = N - 1, \ldots, 0$. These simpler problems may still be hard to solve. The dynamic programming algorithm is feasible only in relatively simple cases, for example if there is only a small finite number of possible values of the state or if the models have relatively simple dynamical and stochastic properties. We give an example from inventory control.

Example 8.2.1 Consider a shop manager who is faced with a randomly varying demand. We assume that the demand can be modelled as a white noise process ε with known probability distribution. The cost of ordering u units, with $u > 0$, is $K + c \cdot u$, with $K \geq 0$ the fixed cost per order and with $c > 0$ the cost per unit. Define $c(u) = K + cu$ for $u > 0$ and $c(u) = 0$ for $u = 0$. Further, let $h \geq 0$ denote the cost of holding inventory per unit and let $p \geq 0$ be the depletion cost per unit demand that can not be met immediately. We assume that this excess demand is fulfilled as soon as additional inventory becomes available and that $p > c$, as else the manager better stops business. Let x denote the stock available at the beginning of a period, then the problem of minimizing the total expected costs over a time horizon of N periods can be formulated as follows. The state evolves according to

$$x(t + 1) = x(t) + u(t) - \varepsilon(t). \tag{8.6}$$

Using the notation $[a]^+ = \max\{0, a\}$, the expected cost is given by

$$J(u) = E \sum_{t=0}^{N-1} \big(c(u(t)) + h[x(t) + u(t) - \varepsilon(t)]^+ + p[\varepsilon(t) - x(t) - u(t)]^+\big). \tag{8.7}$$

The initial stock $x(0)$ is given, and it is assumed that the final stock $x(N)$ has no value.

In order to state the optimal policy it is helpful to define the functions $F(a) = ca + hE[a - \varepsilon]^+ + pE[\varepsilon - a]^+$ and $G_t(a) = F(a) + E\big(J_{t+1}(a - \varepsilon(t))\big)$ with J_{t+1} the optimal-cost-to-go function. Let S be the value minimizing F and let s be the smallest value such that $F(s) = K + F(S)$. Then the single period problem, with $N = 1$, has the following solution. Order $u^*(0) = 0$ if $x(0) \geq s$, and $u^*(0) = S - x(0)$ if $x(0) < s$. This is called the (s, S) policy, below the critical level s one should order an amount so that the target inventory S is reached. If the fixed order costs are zero, that is, $K = 0$, then $s = S$ and this is the desired inventory. If $K = 0$, then the multi period solution also has the form $u^*(t) = 0$ if $x(T) \geq S_t$ and $u^*(t) = S_t - x(t)$ if $x(t) < S_t$, where S_t minimizes G_t. It is somewhat more involved to show that for $K > 0$ the optimal policy is still of the (s_t, S_t) type, with S_t as before and with s_t the smallest value such that $G_t(s_t) = K + G_t(S_t)$. For details we refer to [7].

8.3   LQG Control with State Feedback

The LQG problem is one of the stochastic control problems for which the optimal control law is relatively simple. The *LQG controllers* are popular because the control strategy allows a very straightforward implementation. The definition is as follows.

Definition 8.3.1 The *LQG problem* is the stochastic control problem with f in (8.1) linear, with g_t in (8.3) quadratic, and with $x(0)$ Gaussian and ε in (8.1) Gaussian white noise. Allowable controls are of the form $u(t) = u\{y(s), s \leq t\}$, with y a linear function of x and ε.

We restrict the attention to the time-invariant case, but the results are similar if the parameters change over time. Under the above conditions, the system can be represented as

$$x(t+1) = Ax(t) + Bu(t) + F\varepsilon(t), \tag{8.8}$$

$$y(t) = Cx(t) + G\varepsilon(t), \tag{8.9}$$

and the cost functions as

$$g_N = x^T(N)Q_N x(N) \tag{8.10}$$

$$g_t = \|Rx(t) + Su(t)\|^2 = \left(Rx(t) + Su(t)\right)^T \left(Rx(t) + Su(t)\right). \tag{8.11}$$

So the system is of the ARMAX type with $D = 0$, so that there is no direct feed through from the inputs to the outputs. All matrices appearing in (8.8)–(8.11) are supposed to be known. We will assume throughout that $\varepsilon(t) \sim N(0, I)$ and that $x(0)$ is a Gaussian random variable, $x(0) \sim N(m_0, P_0)$, independent of $\{\varepsilon(t); t \geq 0\}$. This also implies that $x(t)$ and $\varepsilon(s)$ are independent for all $s \geq t$. Further we assume that $S^T S$ is positive definite, so that no controls are without cost. The cost function (8.11) contains as a special case $g_t = y(t)^T Q_1 y(t) + u(t)^T Q_2 u(t)$, with Q_1 and Q_2 positive semidefinite matrices.

Indeed, as $E(x(t)\varepsilon(t)^T) = 0$ it follows that in this case $E(g_t) = E(\widetilde{g_t}) + c$, where $\widetilde{g_t} = x(t)^T C^T Q_1 C x(t) + u(t)^T Q_2 u(t)$ is of the form (8.11) with $R = \begin{pmatrix} Q_1^{1/2} \\ 0 \end{pmatrix}$ and $S = \begin{pmatrix} 0 \\ Q_2^{1/2} \end{pmatrix}$, and where $c = \text{trace}(G^T Q_1 G)$ is independent of the control so that it can be neglected.

In this section we consider the LQG problem with full state observation, that is, we assume that $G = 0, C = I$. Note that in this case $\widetilde{g_t} = g_t$. The general case where the available information consists of the observed outputs $\{y(s), 0 \le s \le t - 1\}$ is discussed in the next section.

For ease of exposition we first summarize the results obtained in Chap. 5 for LQ control. This corresponds to the LQG problem with $F = 0, G = 0, P_0 = 0$ and $C = I$. According to Theorem 5.3.1, see also (5.8) to see the connection, the solution is given by $J_t = x(t)^T Q(t) x(t)$ achieved by the optimal control law

$$u^*(t) = -L(t)x(t), \tag{8.12}$$

where

$$L(t) = \left(B^T Q(t+1)B + S^T S\right)^{-1}\left(B^T Q(t+1)A + S^T R\right), \tag{8.13}$$

and where the matrices Q are generated by the Riccati difference equation

$$Q(t) = A^T Q(t+1)A + R^T R +$$
$$- \left(B^T Q(t+1)A + S^T R\right)^T \left(B^T Q(t+1)B + S^T S\right)^{-1}\left(B^T Q(t+1)A + S^T R\right). \tag{8.14}$$

solved backwards in time starting from the final value

$$Q(N) = Q_N.$$

The infinite horizon problem, with $N \to \infty$, has a solution with finite cost if the pair (A, B) is stabilizable and the pair $(A - B(S^T S)^{-1} S^T R, (I - S(S^T S)^{-1} S^T) R)$ is detectable, for example, if A is stable and $S^T R = 0$ so that the cost function does not involve the cross product between $x(t)$ and $u(t)$. In this case the infinite horizon optimal control law becomes time invariant, that is, $u^*(t) = -Lx(t)$ with $L = (B^T QB + S^T S)^{-1}(B^T QA + S^T R)$ and where Q is the largest Hermitian solution of the following algebraic Riccati equation

$$Q = A^T QA + R^T R - (B^T QA + S^T R)^T (B^T QB + S^T S)^{-1}(B^T QA + S^T R) \tag{8.15}$$

which is positive semidefinite in this case. Also, the closed loop system is stable, that is, $A - BL$ is a stable matrix.

Now we consider the LQG problem with full information on the state. This means that the allowable control strategies are of the form $u(t) = u\{x(s), s \leq t\}$. The following result states that in this case the optimal control law is precisely the same as that for the deterministic LQ problem.

Theorem 8.3.2 *Consider the LQG problem with full state observation, that is, with $G = 0$, and $C = I$ in (8.9), and where ε is Gaussian standard white noise and where $x(0) \sim N(m_0, P_0)$ is independent of $\{\varepsilon(t), t \geq 0\}$. This LQG problem with full state observation has the LQ solution given in (8.12)–(8.14) with minimal cost equal to*

$$E\left(x_0^T Q(0)x_0\right) + \sum_{t=0}^{N-1} \operatorname{trace}(F^T Q(t+1)F) =$$

$$= m_0^T Q(0)m_0 + \operatorname{trace}(Q(0)P_0) + \sum_{t=0}^{N-1} \operatorname{trace}(F^T Q(t+1)F).$$

Proof   The optimal cost-to-go functions are given by

$$J_N(x(N)) = x(N)^T Q_N x(N)$$

and

$$J_t(x(t)) = \inf_{u(t)} \left(\|Rx(t) + Su(t)\|^2 + \right.$$

$$\left. + E\left(J_{t+1}(Ax(t) + Bu(t) + F\varepsilon(t))|x(t)\right)\right).$$

By induction we will prove that $J_t(x(t)) = x(t)^T Q(t)x(t) + c(t)$, with $Q(t)$ as defined in (8.14) and with

$$c(t) = \sum_{k=t}^{N-1} \operatorname{trace}(F^T Q(k+1)F).$$

For $t = N$ this is evident. Suppose it is correct for J_{t+1}, then J_t involves the term

$$E\left(x(t+1)^T Q(t+1)x(t+1) + c(t+1)|x(t)\right),$$

and as $\varepsilon(t)$ is independent of $x(t)$ and $u(t)$ this term is equal to

$$\big(Ax(t) + Bu(t)\big)^T Q(t+1)\big(Ax(t) + Bu(t)\big) +$$
$$+ E\big(\varepsilon(t)^T F^T Q(t+1)F\varepsilon(t)\big) + c(t+1) =$$
$$= \big(Ax(t) + Bu(t)\big)^T Q(t+1)\big(Ax(t) + Bu(t)\big) + c(t).$$

This implies that

$$J_t(x(t)) = \inf_{u(t)} \big(\|Rx(t) + Su(t)\|^2 +$$
$$+ \big(Ax(t) + Bu(t)\big)^T Q(t+1)\big(Ax(t) + Bu(t)\big)\big) + c(t).$$

Up to the constant term $c(t)$, this is precisely the optimal-cost-to-go function of the deterministic LQ problem. It follows from Theorem 5.3.1 that $J_t(x(t)) = x(t)^T Q(t)x(t) + c(t)$, with $Q(t)$ as defined in (8.14). The optimal value of the control input is then of course also the same as in the LQ case, so that (8.12) gives the solution. The optimal cost is $J_0(x(0)) = E(x(0)^T Q(0)x(0)) + c(0)$. Since $x_0 \sim N(m_0, P_0)$ this is easily computed to be $J_0(x(0)) = m_0^T Q(0)m_0 + \text{trace}(Q(0)P_0) + c(0)$.   □

This result shows that in general the costs are unbounded for the infinite horizon problem, as the second cost term will be $\sum_{t=0}^{\infty} \text{trace}(F^T Q F)$ with Q the solution of (8.15). In the deterministic case the costs remain bounded if the control inputs can force the state to zero. In the stochastic case this is not possible, because the disturbances ε in (8.8) will always excite the state. Of course it makes no sense to compare control strategies if even the optimal cost is infinite. In this situation the definition of the cost criterion should be adjusted to obtain suitable comparisons.

One of the possibilities is to consider the discounted cost criterion

$$J_\rho(u) = E\big(\rho^N x^T(N)Q_N x(N) + \sum_{t=0}^{N-1} \rho^t \|Rx(t) + Su(t)\|^2\big) \tag{8.16}$$

where $0 < \rho < 1$ denotes the discount factor. Redefining the state and input as $x_\rho(t) = \rho^{\frac{1}{2}t} x(t)$ and $u_\rho(t) = \rho^{\frac{1}{2}t} u(t)$, it is easily seen that this control problem is equivalent to an undiscounted control problem by redefining (A, B) as $(\rho^{\frac{1}{2}} A, \rho^{\frac{1}{2}} B)$. Note in particular that for ρ sufficiently small the stabilizability and detectability conditions stated before will be satisfied, and then there is a finite optimal discounted cost for the infinite horizon problem. For the discounted cost criterion (8.16), the infinite horizon problem has finite cost if ρ is sufficiently small. Then the optimal control law is of the form (8.12), in terms of the adjusted parameters $(\rho^{\frac{1}{2}} A, \rho^{\frac{1}{2}} B, \rho^{\frac{1}{2}(t+1)} F, \rho^{-\frac{1}{2}t} C)$ instead of (A, B, F, C). The optimal cost is $x_0^T Q_\rho x_0 + \frac{\rho}{1-\rho}\text{trace}(F^T Q_\rho F)$, with Q_ρ the corresponding solution of (8.15).

In a sense, the discounted cost criterion favours the short run performance, as the long run costs get a relatively smaller weight. Another method to obtain finite costs is to consider the long run average cost, defined as

$$\bar{J}(u) = \lim_{N \to \infty} \frac{1}{N} E \sum_{t=0}^{N-1} \| Rx(t) + Su(t) \|^2. \tag{8.17}$$

For a finite horizon the average cost criterium is of course equivalent to the undiscounted total cost criterium, as they only differ by the factor $\frac{1}{N}$. Under the stability and detectability conditions stated before, the optimal control law therefore converges for $N \to \infty$ to the time invariant LQ control law. The result in Theorem 8.3.2 shows that the minimal cost is equal to trace($F^T Q F$) where Q is the solution of (8.15).

8.4   LQG Control with Output Feedback

In the foregoing section we assumed that $y(t) = x(t)$ in (8.9), so that the state is observed. Now we consider the LQG problem in its general form, with observed outputs. So there is only partial information on the state. We first restrict the attention to control laws of the form

$$u(t) = u\{y(s), s \le t - 1\}. \tag{8.18}$$

This has the interpretation that decisions $u(t)$ are made at the beginning of time period t, when the observations $y(t)$ over that period are not yet available. Later we will consider the situation where the control input may also depend on the current output.

We recall that the state is merely an auxiliary variable used to simplify the description of the dynamical relationships between inputs, outputs and disturbances. The complication of the current LQG problem is that the state process is not observed. This would suggest to construct an alternative realization of the system for which the state is observable. Such a realization is obtained by the Kalman filter, because the state $\hat{x}(t) = E(x(t)|y(s), s \le t - 1)$ is a function of past observations. So, by using this alternative state space model we could solve the LQG problem by state feedback as in Theorem 8.3.2. The only complication is that the Kalman filter was derived in Chap. 7 under the assumption that the inputs are exogenous in the sense that $E(u(t)y(s)^T) = 0$ for all t, s, whereas in the current situation the inputs $u(t) = -L(t)\hat{x}(t)$ depend on the past outputs.

We will now first prove that for Gaussian systems the conditional expectation $\hat{x}(t)$ is still generated by the Kalman filter, independent of the control law. This means that the LQG problem has the property of neutrality, because the information on the state process is not influenced by the chosen control action. The assumption of Gaussian distributions is crucial here. It can be shown that neutrality in general does not hold true in the non-

Gaussian case, and that then the best linear predictor of the state under linear control laws need not be given by the Kalman filter.

Theorem 8.4.1 *Consider the system (8.8), (8.9), where G has full row rank, ε is Gaussian standard white noise and where $x(0) \sim N(m_0, P_0)$ independent of $\{\varepsilon(t), t \geq 0\}$. Then for every control law $u(t) = u\{y(s); s \leq t\}$, the conditional expectation $\hat{x}(t) = E(x(t)|y(s), s \leq t-1)$ and its covariance $P(t) = E\big(x(t)-\hat{x}(t)\big)\big(x(t)-\hat{x}(t)\big)^T$ are given by the Kalman filter (7.12)–(7.14), with starting conditions $\hat{x}(0) = m_0$ and $P(0) = P_0$. The processes u, y, x, and $\hat{x}$ are in general not Gaussian, but the innovations process $\omega = y - C\hat{x}$ is Gaussian.*

Proof The control law need not be linear, and therefore u, y, x, and $\hat{x}$ are in general not Gaussian. (Note that they are Gaussian if the control law is linear.)

The idea is to split the system in two parts, one subsystem describing the effect of the control inputs and the other one the effect of the disturbances. We indicate the controlled part by a subindex c and the noisy part by a subindex n. So let

$$x_c(t + 1) = Ax_c(t) + Bu(t),$$

$$x_c(0) = m_0,$$

$$y_c(t) = Cx_c(t)$$

and let

$$x_n(t + 1) = Ax_n(t) + F\varepsilon(t),$$

$$x_n(0) = x(0) - m_0 \sim N(0, P_0),$$

$$y_n(t) = Cx_n(t) + G\varepsilon(t).$$

Then $x = x_c + x_n$ and $y = y_c + y_n$. An according decomposition of the predictor is given by $\hat{x} = \hat{x}_c + \hat{x}_n$, where $\hat{x}_c(t) = E(x_c(t)|y(s), s \leq t-1)$ and $\hat{x}_n(t) = E(x_n(t)|y(s), s \leq t-1)$. In fact there holds that

$$\hat{x}(t) = x_c(t) + E\left(x_n(t)|y_n(s), s \leq t - 1\right). \tag{8.19}$$

To prove this, note that $x_c(t)$ is a function of $\{u(s), s \leq t - 1\}$ as $x_c(0) = m_0$ is known, so it is also a function of $\{y(s), s \leq t - 1\}$ and hence $\hat{x}_c = x_c$. Concerning the noisy part, it is sufficient to prove that $\{y(s), s \leq t - 1\}$ and $\{y_n(s), s \leq t - 1\}$ contain the same information, that is, that there exists a bijection between these two sets of random variables.

To see this, notice that $y_n(s) = y(s) - y_c(s) = y(s) - Cx_c(s)$ is a function of the known initial condition m_0 and of $\{y(s), u(r), r \leq s - 1\}$, hence of $\{y(r), r \leq s\}$.

Conversely, $y(0) = y_n(0) + Cm_0$, and supposing that for all $\sigma \leq s$, $y(\sigma)$ is a function of $\{y_n(r), r \leq \sigma\}$ it follows that $y(s+1) = y_n(s+1) + C(Ax_c(s) + Bu(s))$ is a function of $\{y_n(s+1), u(r), r \leq s\}$, hence of $\{y_n(s+1), y(r), r \leq s\}$, and by the induction assumption hence also of $\{y_n(r), r \leq s+1\}$. This proves (8.19).

In (8.19) x_c is deterministic and $\hat{x}_n$ is independent of the control inputs. Therefore $\hat{x}_n$ can be calculated by the standard Kalman filter, $\hat{x}_n(t+1) = A\hat{x}_n(t) + K(t)\omega_n(t)$ where $\omega_n = y_n - C\hat{x}_n$ with the Kalman gain (7.13), (7.14). As $x_n(0) \sim N(0, P_0)$ the starting conditions are $\hat{x}_n(0) = 0$ with $P(0) = P_0$. The innovation process ω_n is Gaussian, as y_n and $\hat{x}_n$ are Gaussian, and as $\omega = y - C\hat{x} = y_c + y_n - Cx_c - C\hat{x}_n = \omega_n$ the innovation process ω is also Gaussian. Further, from (8.19) and $\omega = \omega_n$ it follows that $\hat{x}(t+1) = x_c(t+1) + \hat{x}_n(t+1) = A\hat{x}(t) + Bu(t) + K(t)\omega(t)$ which coincides with the Kalman filter equation (7.12). The starting condition is $\hat{x}(0) = x_c(0) + \hat{x}_n(0) = m_0$. Finally, $P(t)$ is the state error covariance matrix, as $x_n(t) - \hat{x}_n(t) = x_n(t) + x_c(t) - (\hat{x}_n(t) + x_c(t)) = x(t) - \hat{x}(t)$ so that $P(t) = E(x_n(t) - \hat{x}_n(t))(x_n(t) - \hat{x}_n(t))^T = E(x(t) - \hat{x}(t))(x(t) - \hat{x}(t))^T$. $\square$

Theorem 8.4.1 provides an alternative state space representation of the system (8.8), (8.9) with observed state, namely

$$\hat{x}(t+1) = A\hat{x}(t) + Bu(t) + K(t)\omega(t) \tag{8.20}$$

$$y(t) = C\hat{x}(t) + \omega(t). \tag{8.21}$$

The initial condition is $\hat{x}(0) = m_0$, and ω is a Gaussian white noise process with $\omega(t) \sim N(0, CP(t)C^T + GG^T)$. So the parameters of this state space model are time varying. The optimal control law is now obtained from Theorem 8.3.2.

Theorem 8.4.2 *The optimal LQG controller of the form $u(t) = u\{y(s), s \leq t-1\}$ for the system (8.8), (8.9) and the cost (8.10), (8.11) is given by*

$$u^*(t) = -L(t)\hat{x}(t) \tag{8.22}$$

The feedback gain $L(t)$ is obtained by (8.13) and (8.14), and $\hat{x}(t)$ is obtained by (8.20) and (8.21) with Kalman filter gain (7.13), (7.14). The minimal cost is given by

$$m_0^T Q(0)m_0 + \text{trace}\left(Q(0)P(0) + Q_N P(N)\right) +$$

$$+ \sum_{t=0}^{N-1} \text{trace}\left(RP(t)R^T + Q(t+1)K(t)\left(CP(t)C^T + GG^T\right)K^T(t)\right). \tag{8.23}$$

Proof   Note that we can rewrite (8.20) into the form

$$\hat{x}(t+1) = A\hat{x}(t) + bu(t) + F(t)\epsilon t,$$

with $F(t) = K(t)\big(CP(t)C^t + GG^T\big)^{\frac{1}{2}}$. We shall apply Theorem 8.3.2 to this situation. In order to do so, we first express the cost function in terms of $\hat{x}$ instead of x. For this purpose let $\tilde{x}(t) = x(t) - \hat{x}(t)$. From (8.19) it follows that $\tilde{x}(t) = x_n(t) - \hat{x}_n(t)$, so that $\tilde{x}(t) \sim N(0, P(t))$, which implies $E(\tilde{x}(t)) = \text{trace}(RP(t)R^T$, and $E(\tilde{x}(t)y(s)^T) = 0$ for $s \leq t - 1$. Let E_{t-1} denote the conditional expectation with respect to $\{y(s), s \leq t - 1\}$, then $E_{t-1}\tilde{x}(t) = E_{t-1}(x(t) - \hat{x}(t)) = \hat{x}(t) - \hat{x}(t) = 0$. As $u(t) = u\{y(s), s \leq t - 1\}$ it follows that

$$E_{t-1}\|Rx(t) + Su(t)\|^2 = E_{t-1}\|R\hat{x}(t) + R\tilde{x}(t) + Su(t)\|^2 =$$

$$= \|R\hat{x}(t) + Su(t)\|^2 + E_{t-1}\|R\tilde{x}(t)\|^2 + 2\big(R\hat{x}(t) + Su(t)\big)^T RE_{t-1}\tilde{x}(t) =$$

$$= \|R\hat{x}(t) + Su(t)\|^2 + \text{trace}\left(RP(t)R^T\right).$$

Further for the final cost

$$E_{N-1}\big(x(N)^T Q_N x(N)\big) =$$

$$= \hat{x}(N)^T Q_N \hat{x}(N) + E_{N-1}\big(\tilde{x}(N)^T Q_N \tilde{x}(N)\big) + 2\hat{x}(N)^T Q_N E_{N-1}\tilde{x}(N) =$$

$$= \hat{x}(N)^T Q_N \hat{x}(N) + \text{trace}(Q_N P(N)).$$

Using these results, the cost function can be reformulated as

$$J(u) = E\big(x(N)^T Q_N x(N) + \sum_{t=0}^{N-1} \|Rx(t) + Su(t)\|^2\big) =$$

$$= E\big(E_{N-1}x(N)^T Q_N x(N) + \sum_{t=0}^{N-1} E_{t-1}\|Rx(t) + Su(t)\|^2\big) =$$

$$= E\big(\hat{x}(N)^T Q_N \hat{x}(N) + \sum_{t=0}^{N-1} \|R\hat{x}(t) + Su(t)\|^2\big) +$$

$$+ \text{trace}\big(Q_N P(N)\big) + \sum_{t=0}^{N-1} \text{trace}\big(RP(t)R^T\big). \tag{8.24}$$

So, apart from the constant term (8.24) which does not depend on the control law, the cost function in terms of $\hat{x}$ is precisely the same as the original one in terms of x. The optimal control law is then obtained from Theorem 8.3.2, with $\hat{x}$ replacing x. Actually, in Theorem 8.3.2 all parameters are time-invariant and the white noise ε is distributed as $\varepsilon(t) \sim N(0, I)$, whereas in the current case the parameter K in (8.20) varies over time and the white noise ω has distribution $\omega(t) \sim N(0, CP(t)C^T + GG^T)$. However, Theorem 8.3.2 holds also true for parameters that vary over time, provided that all parameters are known (compare [44, Section 3.6], where this is shown for continuous time systems). This is the case here, see (7.13) and (7.14). It follows that the control law (8.22) is optimal. In order to compute the minimal cost, recall that (8.20) can be standardised by defining $F(t) = K(t)(CP(t)C^T + GG^T)^{\frac{1}{2}}$. Then from the foregoing expression of $J(u)$ and Theorem 8.3.2 we obtain

$$J(u) = E\big(x(0)^T Q(0)x(0)\big) + \sum_{t=0}^{N-1} \text{trace}\big(F(t)^T Q(t+1)F(t)\big) +$$

$$+\text{trace}\big(Q_N P(N)\big) + \sum_{t=0}^{N-1} \text{trace}\big(RP(t)R^T\big),$$

and because $E\big(x(0)^T Q(0)x(0)\big) = m_0^T Q(0)m(0) + \text{trace}\big(Q(0)P_0\big)$ this proves the expression for the minimal cost given in the theorem. $\qquad\square$

The optimal LQG controller has a simple recursive structure. The two involved Riccati difference equations, (7.21) for the filter, and (8.14) for the controller, can be solved off-line, independent of the observations. This gives the Kalman filter gains (7.13) and the feedback gains (8.13). The state is recursively estimated by the Kalman filter (7.12), and then the control is obtained by linear feedback (8.22).

The LQG controller has the certainty equivalence property. It consists of the deterministic LQ controller (8.12), with the unobserved state replaced by its optimal estimate. Further, it also satisfies the separation principle. This means that the form of the controller is independent of the parameters (F, G) of the stochastic part of the model and that the form of the estimator is independent of the parameters (Q, R, S) of the control objectives.

It can be shown that the LQG problem with control law $u(t) = u\{y(s), s \le t\}$ has optimal solution $u^*(t) = -L(t)\hat{x}(t|t)$, where $\hat{x}(t|t) = E(x(t)|y(s), s \le t)$ can be obtained recursively as described in Proposition 7.3.3.

Under the stabilizability and detectability conditions of Theorems 5.3.2 and 7.4.2 the controller and the filter become time-invariant in the limit if the number of observations $N \to \infty$. The state observer is then obtained from (7.22), (7.23), and (7.24), and the controller from (8.13), (8.15). It follows from Theorem 8.4.2 that the long-run average cost (8.17) is in this case equal to

$$\bar{J}(u) = \text{trace}\big(RPR^T + QK(CPC^T + GG^T)K^T\big).$$

For the discounted cost criterion (8.16) the optimal controller is given by $u^*(t) = -L_\rho(t)\hat{x}(t)$, with $\hat{x}(t)$ as before and with the feedback gain calculated in terms of the transformed parameters $(\rho^{\frac{1}{2}}A, \rho^{\frac{1}{2}}B, \rho^{\frac{1}{2}(t+1)}F, \rho^{-\frac{1}{2}}C)$. Under the conditions of Theorems 5.3.2 and 7.4.2 the filter and the controller become again time-invariant for $N \to \infty$.

The controller gain (8.13) is in terms of the solution Q_ρ of (8.15) with the transformed parameters $(\rho^{\frac{1}{2}} A, \rho^{\frac{1}{2}} B)$ instead of (A, B).

In practice it may be difficult to choose the parameters (R, S, Q_N) of the cost function. An alternative is minimum variance control, which uses the criterion (8.17) with $R = C$ and $S = 0$. In this case the costs are expressed in terms of the outputs, and the possible costs of the inputs are neglected. This violates the assumption in LQG that $S^T S$ is positive definite. However, the criterion can be approximated by choosing $S = \alpha I$ with α sufficiently small.

Further, in practice the system parameters (A, B, C, F, G) are often not known. A possible approach is to use the certainty equivalence principle, that is, first estimate the unknown parameters and then determine the control law. The estimation of system parameters from observed data is called the system identification problem. The parameters may be varying over time, and it may then be necessary to update the parameter estimates when new observations become available. This is called recursive identification, and in connection with control applications this is called adaptive control.

We conclude by an example of LQG control with output feedback.

Example 8.4.1 Consider the ARMAX system with single input and single output

$$y(t) = y(t - 1) + u(t - 2) + \varepsilon(t)$$

where ε is a standard white noise process with $E\varepsilon(t)y(s) = 0$ for all $s < t$. The input $u(t)$ is allowed to be a function of the outputs $y(s), s \leq t - 1$; i.e. $u(t) = u\{y(s), s \leq t - 1\}$. As cost function we consider

$$J(u) = \lim_{N \to \infty} \sum_{t=1}^{N} E(y(t)^2 + u(t)^2).$$

Define two state variables by $x_1(t) = y(t - 1) + u(t - 2)$ and $x_2(t) = u(t - 1)$, then a state space realization (8.8), (8.9) is obtained with

$$A = \begin{pmatrix} 1 & 1 \\ 0 & 0 \end{pmatrix}, \quad B = \begin{pmatrix} 0 \\ 1 \end{pmatrix}, \quad C = \begin{pmatrix} 1 & 0 \end{pmatrix}, \quad F = \begin{pmatrix} 1 \\ 0 \end{pmatrix}, \quad G = 1.$$

For controls $u(t) = u\{y(s), s \leq t - 1\}$ it follows that $E(y(t)^2 + u(t)^2) = E(x_1(t)^2 + \varepsilon(t)^2 + 2x_1(t)\varepsilon(t) + u(t)^2) = 1 + E(x_1(t)^2 + u(t)^2)$, so the control objective is to minimize (8.17) with

$$R = \begin{pmatrix} 1 & 0 \\ 0 & 0 \end{pmatrix}, \quad S = \begin{pmatrix} 0 \\ 1 \end{pmatrix}.$$

Concerning the conditions in Theorem 5.3.2 there holds $S^T R = 0$ and (A, B) is controllable (hence stabilizable) and (A, R) is observable (hence detectable). Also the conditions of Theorem 7.4.2 are satisfied, because (A, C) is observable and $A - FG^T (GG^T)^{-1} C = \begin{pmatrix} 0 & 1 \\ 0 & 0 \end{pmatrix}$ is a stable matrix. From Theorem 8.4.2 it follows that the optimal control law is given by the feedback $u(t) = -L\hat{x}(t)$ where $L = (B^T QB + S^T S)^{-1} B^T QA$, see (8.13), (8.14), and (8.15) and with $\hat{x}$ the filtered state in (7.22), (7.23), and (7.24). As the two state variables in our model are functions of past inputs and outputs, it follows that $\hat{x} = x$. This can also be checked by observing that the filter equation (7.22) has solution $P = 0$. This means in (7.24) that $K = \begin{pmatrix} 1 \\ 0 \end{pmatrix}$. Direct calculation of the solution $Q = \begin{pmatrix} q_1 & q_2 \\ q_2 & q_3 \end{pmatrix}$ of the control equation (8.15) shows that $q_1 = \frac{1}{2}(3 + \sqrt{5})$ and $q_2 = q_3 = \frac{1}{2}(1 + \sqrt{5})$, so that the feedback gain is given by $L = \begin{pmatrix} f & f \end{pmatrix}$, where $f = \frac{q_2}{(q_3+1)}$. Substituting the expressions for the state variables, this gives the control law

$$u^*(t) = -f\big(y(t - 1) + u(t - 1) + u(t - 2)\big).$$

With this control law, the closed loop system is

$$y(t) = -fy(t - 1) + \varepsilon(t) + f\varepsilon(t - 1) + f\varepsilon(t - 2).$$

Without controls, that is, with $u(t) = 0$ for all $t \in \mathbb{Z}$, the output process y is not stationary and $Ey(t)^2 \to \infty$ for $t \to \infty$. However, the controlled system makes the process y stationary because $|f| < 1$ in the above ARMA(1,2) model.

System Identification

9

System identification is concerned with the estimation of a system on the basis of observed data. This involves specification of the model structure, estimation of the unknown model parameters, and validation of the resulting model. Least squares and maximum likelihood methods are discussed, for stationary processes (without inputs) and for input-output systems.

9.1   Identification

In the foregoing chapters we always assumed that the system is known to us, and we considered the representation, regulation, and prediction of linear systems with given parameters. In most practical applications the system is not known and has to be estimated from the available information. This is called the identification problem. The identification method will depend on the intended model use, as this determines what aspects of the system are of relevance. The three main choices in system identification are the following.

(i) *Data* In some situations it is possible to generate a large amount of reliable data by carefully designed experiments. In other situations the possibilities to obtain data are much more limited and it is not possible to control for external factors that influence the outcomes. That is, the magnitude of outside disturbances ('noise') may differ widely from one application to another.

(ii) *Model Class* A model describes relations between the observed variables. For practical purposes the less important aspects are neglected to obtain sufficiently simple models. The identified model should be validated to test whether the imposed simplifications are acceptable.

© The Author(s), under exclusive license to Springer Nature Switzerland AG 2021
C. Heij et al., *Introduction to Mathematical Systems Theory*,
https://doi.org/10.1007/978-3-030-59654-5_9

(iii) *Criterion* The criterion reflects the objectives of the modeller. It expresses the usefulness of models in representing the observed data.

In practice, system identification often involves several runs of the empirical cycle which consists of the specification of the problem, the estimation of a model by optimization of the criterion, the validation of the resulting model, and possible adjustments that may follow from this validation.

In the following we restrict our attention to linear systems, quadratic criteria and data that consists of observed time series of the system variables. The advantage of this linear quadratic framework is that it leads to relatively simple identification algorithms. Further, the ideas and concepts for these methods form the basis for more advanced approaches.

Models are simplifications of reality and therefore they involve errors. It is often assumed that the data can be decomposed into two parts, a systematic part (related to the underlying system) and a disturbance part that reflects unmodelled aspects of the system. By assuming that the disturbances are random variables, the statistical properties of identification methods can be evaluated. In particular, one considers the properties of unbiasedness, efficiency, and consistency. Let θ denote the unknown system parameters, and let $\hat{\theta}$ be an estimator of θ based on the observed data. Because the data are influenced by the random disturbances, the estimator $\hat{\theta}$ is also a random variable. It is called an *unbiased estimator* if $E(\hat{\theta}) = \theta$, and it is called an *efficient estimator* in a class of estimators if it minimizes the variance $\text{var}(\hat{\theta}) = E(\hat{\theta} - E(\hat{\theta}))(\hat{\theta} - E(\hat{\theta}))^T$, that is, if for every other estimator $\tilde{\theta}$ in this class $\text{var}(\tilde{\theta}) - \text{var}(\hat{\theta})$ is a positive semidefinite matrix. To define consistency, let $\hat{\theta}_N$ denote the estimator based on data that are observed on a time interval of length N. The estimator is called (*weakly*) *consistent* if, for every $\delta > 0$, there holds

$$\lim_{N \to \infty} P(\|\hat{\theta}_N - \theta\| \geq \delta) = 0 \tag{9.1}$$

where $\| \cdot \|$ denotes the Euclidean norm. This is also written as $\text{plim}(\hat{\theta}_N) = \theta$. Hereby it is assumed that the system under investigation belongs to the model class, but this can be generalized to the situation where θ is the optimal (but not perfectly correct) model within the model class.

9.2 Regression Models

In this section we consider single input, single output systems with a finite impulse response (FIR), that is,

$$y(t) = \beta_1 u(t-1) + \cdots + \beta_k u(t-k) + \varepsilon(t) \tag{9.2}$$

We assume that y is observed for $t = 1, \ldots, N$, and u for $t = 1 - k, \ldots, N - 1$ with $N \geq k$. Let $x(t) := \big(u(t - 1), \ldots, u(t - k)\big)^T$ and let $y = \big(y(1), \cdots, y(N)\big)^T$, $X = \big(x(1), \ldots, x\big)^T$, $\varepsilon = \big(\varepsilon(1), \ldots, \varepsilon(N)\big)^T$ and $\beta = (\beta_1, \ldots, \beta_k)^T$. Then (9.2) can be written as the regression model

$$y = X\beta + \varepsilon. \tag{9.3}$$

In the sequel, whenever necessary, we shall write X_N instead of X to emphasize the dependence of X on N.

From the data, y and X, we have to estimate the parameters β. The least squares estimator $\hat{\beta}$ minimizes the sum of squared errors

$$\sum_{t=1}^{N} \varepsilon^2(t) = \|\varepsilon\|^2 = \|y - X\beta\|^2.$$

This is obtained by projecting y onto the column space of X, so that $X^T(y - X\hat{\beta}) = 0$. Assuming that $\text{rank}(X) = k$, the solution is given by

$$\hat{\beta} = (X^T X)^{-1} X^T y \tag{9.4}$$

In order to investigate under which conditions this is a good estimator, we make the following assumptions.

Assumptions
The data satisfy the relation $y = X\beta + \varepsilon$, where

A1   all entries of the matrix X are non-random, and $\text{rank}(X) = k$;
A2   all entries of the (unobserved) disturbance vector ε are outcomes of random variables with $E(\varepsilon) = 0$, $E(\varepsilon^2(t)) = \sigma^2$ (equal variance), and $E(\varepsilon(t)\varepsilon(s)) = 0$ for all $t \neq s$ (no serial correlation).

Definition 9.2.1   We call an estimator *linear* if it is of the form $\tilde{\beta} = Ay$, with A a non-random matrix, and it is called a *best linear unbiased estimator* (BLUE) if it is unbiased with minimal variance in the class of all linear unbiased estimators.

The following result is called the *Gauss-Markov theorem*.

Theorem 9.2.2   *Under assumptions A1 and A2, the least squares estimator* (9.4) *is BLUE with* $\text{var}(\hat{\beta}) = \sigma^2(X^T X)^{-1}$. *A sufficient condition for consistency is that*

$$\lim_{N \to \infty} \lambda_{min}(X_N^T X_N) = \infty,$$

where X_N is the regressor matrix in (9.3) for the first N observations and λ_{min} denotes the smallest eigenvalue.

Proof It follows from (9.3) and (9.4) that $\hat{\beta} = \beta + (X^T X)^{-1} X^T \varepsilon$. As X is non-random, $E(\varepsilon) = 0$ and $\text{var}(\varepsilon) = \sigma^2 I$, it follows that $E(\hat{\beta}) = \beta$ and

$$\text{var}(\hat{\beta}) = (X^T X)^{-1} X^T \text{var}(\varepsilon) X (X^T X)^{-1} = \sigma^2 (X^T X)^{-1}.$$

Let $\tilde{\beta} = Ay$ be another unbiased estimator and define $\Delta = A - (X^T X)^{-1} X^T$. Unbiasedness requires that $E(\tilde{\beta}) = AX\beta = \beta$ for every β, so that $AX = I$ and $\Delta X = 0$. Then $\tilde{\beta} - E\tilde{\beta} = A(X\beta + \varepsilon) - \beta = A\varepsilon$ and

$$\text{var}(\tilde{\beta}) = E(\tilde{\beta} - E\tilde{\beta})(\tilde{\beta} - E\tilde{\beta})^T = E(A\varepsilon\varepsilon^T A^T) = \sigma^2 A A^T$$
$$= \sigma^2 (\Delta\Delta^T + (X^T X)^{-1}) = \sigma^2 \Delta\Delta^T + \text{var}(\hat{\beta}).$$

As $\Delta\Delta^T$ is positive semidefinite this shows that $\hat{\beta}$ is BLUE.

From now on we emphasize that $X = X_N$ and denote $\hat{\beta}$ by $\hat{\beta}_N$.

To prove consistency we use the Markov inequality, that is, for every random variable z and every $c > 0$ there holds $E(z^2) \geq c^2 P(|z| \geq c)$ so that $P(|z| \geq c) \leq c^{-2} E(z^2)$. It then follows that for every $\delta > 0$

$$P(\|\hat{\beta}_N - \beta\| \geq \delta) \leq P(|\hat{\beta}_{n,i} - \beta_i| \geq k^{-\frac{1}{2}}\delta \text{ for some } i = 1, \cdots, k) \leq$$
$$\leq k\delta^{-2} E(\hat{\beta}_{N,i} - \beta_i)^2 = k\delta^{-2}\text{var}(\hat{\beta}_{N,i}) \leq k\delta^{-2}\sigma^2 \lambda_{max}\{(X_N^T X_N)^{-1}\} =$$
$$= k\delta^{-2}\sigma^2\{\lambda_{min}(X_N^T X_N)\}^{-1}$$

and this converges to zero for $N \to \infty$, by assumption.   □

Returning to the FIR system (9.2), assumptions A1 and A2 mean that the input is not random but the output is random. This may be relevant in experimental situations where the input is controlled. However, often the input will be affected by uncertain factors that fall outside the scope of the model. The above results remain asymptotically valid for random inputs, provided some conditions are satisfied. We restrict the attention to consistency, and replace assumption A1 by the following.

A1* The matrix X is random and such that $\text{plim}(\frac{1}{N}X_N^T X_N) = Q$ exists with Q invertible (sufficiency of excitation).

For the FIR system (9.2) there holds $\frac{1}{N}X_N^T X_N = \frac{1}{N}\sum_{t=1}^{N} x(t)^T x(t)$, where $x(t) = (u(t-1), \ldots, u(t-k))$, so that Q corresponds to the covariance matrix of the input and its lags. The excitation condition basically means that the input satisfies no polynomial equations and that it does not die out when $N \to \infty$.

Theorem 9.2.3 *Under assumptions A1* and A2, the least squares estimator is consistent if and only if* $\text{plim}(\frac{1}{N}\sum_{t=1}^{N} x(t)^T \varepsilon(t)) = 0$ *(orthogonality condition).*

Proof The least squares estimator is $\hat{\beta}_N = \beta + (\frac{1}{N}X_N^T X_N)^{-1}(\frac{1}{N}X_N^T \varepsilon)$ where $\frac{1}{N}X_N^T \varepsilon = \frac{1}{N}\sum_{t=1}^{N} x(t)^T \varepsilon(t)$. The definition of convergence in probability gives that if $\text{plim}(a_n) = a$ and f is a continuous function, then $\text{plim}(f(a_n)) = f(a)$. Therefore $\text{plim}(\hat{\beta}_N) = \beta + Q^{-1}\text{plim}(\frac{1}{N}X_N^T \varepsilon)$, which proves the result. □

The orthogonality condition essentially requires that the regressor variables $x(t)$ show no contemporaneous correlation with the error term $\varepsilon(t)$. For the FIR system this means that the output error in (9.2) is uncorrelated with the past inputs.

Many time series that are observed in practice show trends and seasonal variation. The modelling of trends and seasonals is discussed in the next chapter. In the current chapter we will either assume that the data are stationary, which can sometimes be achieved by appropriate data transformations, or that the model explicitly includes variables for the nonstationary part.

9.3 Maximum Likelihood

Stochastic models assign (relative) probabilities to the observations of the system variables. Suppose that the model class consists of a set of probability densities $\{p_\theta, \theta \in \Theta\}$, where $\theta \in \Theta$ is the vector of unknown parameters. If the data consists of q time series that are observed on a time interval of length N, then p_θ is a probability density on $(\mathbb{R}^q)^N$. The maximum likelihood method chooses the model that assigns the highest probability to the observed data. If we denote the data by $w \in (\mathbb{R}^q)^N$, then this means that the likelihood function $L(\theta) := p_\theta(w)$ is maximized over the parameter set Θ.

Maximum likelihood estimation (ML) requires that the probability distribution is specified as an explicit function of the parameters θ. As an example, we consider the regression model (9.3) $y = X\beta + \varepsilon$. In this case, the parameters θ are given by $(\beta^T, \sigma^2)^T$. We extend assumption A2 as follows.

A2* The disturbance vector ε has the multivariate normal distribution with mean $E(\varepsilon) = 0$ and covariance matrix $E(\varepsilon\varepsilon^T) = \sigma^2 I$.

Theorem 9.3.1 *Under assumptions A1 and A2*, the maximum likelihood estimators in the regression model (9.3) are given by* $\hat{\beta} = (X^T X)^{-1} X^T y$ *and* $\hat{\sigma}^2 = \frac{1}{N}(y - X\hat{\beta})^T (y - X\hat{\beta})$.

Proof Let $\theta = (\beta^T, \sigma^2)^T$ denote the vector of the model parameters. As $\varepsilon = y - X\beta$ has the normal distribution, the likelihood function is given by

$$L(\beta, \sigma^2) = p_\theta(y, X) = (2\pi\sigma^2)^{-\frac{N}{2}} \exp\{-(2\sigma^2)^{-1}(y - X\beta)^T (y - X\beta)\} \tag{9.5}$$

As the logarithm is a monotonic function, maximization of $L(\beta, \sigma^2)$ is equivalent to maximization of

$$\frac{2}{N} \log L(\beta, \sigma^2) = -\log(2\pi) - \log(\sigma^2) - \frac{1}{2\sigma^2} \frac{1}{N} (y - X\beta)^T (y - X\beta).$$

It follows that the maximum is obtained for $\hat{\beta} = (X^T X)^{-1} X^T y$ and that $\hat{\sigma}^2 = \frac{1}{N}(y - X\hat{\beta})^T (y - X\hat{\beta})$. □

Theorem 9.3.2 *Under assumptions A1 and A2*, the least squares estimator $\hat{\beta}$ in (9.4) is minimum variance unbiased, that is, it is unbiased and if $\tilde{\beta}$ is another unbiased estimator then* $\text{var}(\tilde{\beta}) - \text{var}(\hat{\beta})$ *is positive semidefinite.*

Proof   Again, let $\theta = (\beta^T, \sigma^2)^T$ denote the model parameters. The Cramer-Rao theorem states that every unbiased estimator $\hat{\theta}$ has a covariance matrix $\text{var}(\hat{\theta}) \geq [-E(\frac{\partial^2 \log L}{\partial \theta \partial \theta^T})]^{-1}$, see [40]. It follows by direct calculation from (9.5) that in this case the lower bound is a block-diagonal matrix with blocks $\sigma^{-2}(X^T X)$ and $(2\sigma^4)^{-1} N$. This implies that for every unbiased estimator there holds $\text{var}(\tilde{\beta}) \geq \sigma^2 (X^T X)^{-1} = \text{var}(\hat{\beta})$, see Theorem 9.2.2.   □

Under very general conditions, maximum likelihood estimators have optimal asymptotic properties, provided that the model is correctly specified. That is, if the data are generated by a probability distribution p_{θ^0}, with $\theta^0 \in \Theta$, and $\hat{\theta}_N$ is the ML estimate based on N observations, then under very general conditions there holds that

   (i) $\hat{\theta}_N$ is consistent, that is, $\text{plim}(\hat{\theta}_N) = \theta^0$;
  (ii) $\hat{\theta}_N$ is asymptotically efficient in the class of all consistent estimators, that is, $\lim_{N \to \infty} N(\text{var}(\tilde{\theta}_N) - \text{var}(\hat{\theta}_N))$ is positive semidefinite for every consistent estimator $\tilde{\theta}$;
 (iii) $\hat{\theta}_N$ has an asymptotic normal distribution, in the sense that $\sqrt{N}(\hat{\theta}_N - \theta^0)$ converges to a normal distribution with mean zero and covariance matrix $[-E(\frac{\partial^2 \log L}{\partial \theta \partial \theta^T})]^{-1}$.

We refer to , e.g., [25] for a proof of this result. From a computational point of view, ML estimation requires the maximization of the likelihood function or equivalently, of its logarithm, both of which are functions of several real variables. The first order conditions will in general consist of a set of nonlinear equations in θ that can be solved by numerical methods. Such methods differ in the choice of initial estimates, search strategies, and convergence criteria. The Newton-Raphson method consists of an iterative linearization of the stationarity condition for a maximum. Consider this for the maximization of the logarithm of the likelihood functions. If $\hat{\theta}_i$ is the current estimate, $G_i = \frac{\partial \log L(\theta)}{\partial \theta}$ the gradient and $H_i = \frac{\partial^2 \log L}{\partial \theta \partial \theta^T}$ the Hessian in $\hat{\theta}_i$, then locally around $\hat{\theta}_i$ there holds

$\frac{\partial \log L(\theta)}{\partial \theta} \approx G_i + H_i(\theta - \hat{\theta}_i)$ by Taylor's formula. This motivates the iterations

$$\hat{\theta}_{i+1} = \hat{\theta}_i - H_i^{-1} G_i \tag{9.6}$$

A possible disadvantage is that this requires the computation and inversion of the Hessian matrix. For nonlinear regression models of the form

$$y(t) = f(x(t), \theta) + \varepsilon(t) \tag{9.7}$$

one could use the Gauss-Newton method for the minimization of $\sum_{t=1}^{N} \varepsilon^2(t)$ as an alternative. This corresponds to maximum likelihood if the disturbances satisfy assumption A2*. Here $x(t)$ is the vector of regressors at time t, and f is a nonlinear function of the model parameters θ. If $\hat{\theta}_i$ is the current estimate, then the model (9.7) is linearized by $f(x, \theta) \approx f(x, \hat{\theta}_i) + x_i^T(\theta - \hat{\theta}_i)$, where $x_i = \frac{\partial}{\partial \theta} f(x, \theta)$ is the gradient evaluated at $(x, \hat{\theta}_i)$. The linearized model gives $\varepsilon(t) = y(t) - f(x(t), \theta) \approx y(t) - f(x(t), \hat{\theta}_i) - x_i^T(t)(\theta - \hat{\theta}_i) = \varepsilon_i(t) - x_i^T(t)(\theta - \hat{\theta}_i)$, where $\varepsilon_i(t)$ denotes the residuals of (9.7) for the estimate $\hat{\theta}_i$ and $x_i(t)$ is the gradient of f at $(x(t), \hat{\theta}_i)$. The corresponding approximation of the criterion function gives $\sum_{t=1}^{N} \varepsilon^2(t) \approx \sum_{t=1}^{N} \{\varepsilon_i(t) - x_i^T(t)(\theta - \hat{\theta}_i)\}^2$. This is a least squares problem with estimate $\hat{\theta}_{i+1} = (X_i^T X_i)^{-1} X_i^T (\varepsilon_i + X_i \hat{\theta}_i)$, that is

$$\hat{\theta}_{i+1} = \hat{\theta}_i + (X_i^T X_i)^{-1} X_i^T \varepsilon_i \tag{9.8}$$

Here X_i is the matrix with N rows consisting of the gradients $x_i(t)$, $t = 1, \cdots, N$, and ε_i is the $N \times 1$ vector with the residuals for $\hat{\theta}_i$.

9.4 Estimation of Autoregressive Models

In this section, we suppose that the data consists of observations of a single output variable $y(t)$, observed for $t = 1, \cdots, N$, and generated by an autoregressive model

$$y(t) = \alpha_1 y(t-1) + \cdots + \alpha_p y(t-p) + \varepsilon(t). \tag{9.9}$$

Here ε is a white noise process with mean zero, variance σ^2, and finite fourth order moments, so that assumption A2 is satisfied. We assume that this model is causal, that is, that the polynomial $1 - \sum_{i=1}^{p} \alpha_i z^{-i}$ has all its roots inside the unit disc. Moreover, we assume that p is known and correctly specified. In Sect. 9.6.1 we shall discuss methods to estimate the lag order p from the data.

Theorem 9.4.1 *The least squares estimator of $(\alpha_1, \cdots, \alpha_p)$ in a causal autoregressive model (9.9) is consistent.*

Outline of Proof According to Theorem 9.2.3, it suffices to prove that assumption A1* is satisfied and that $\text{plim}(\frac{1}{N}\sum_{t=1}^{N}\varepsilon(t)y(t-i)) = 0$ for $i = 1, \cdots, p$. As was discussed in Sect. 6.3, stationarity implies that $y(t)$ can be written as a function of the past disturbances $\{\varepsilon(s), s \leq t\}$. Therefore $E(\varepsilon(t)y(t-i)) = 0$ for all t and $i = 1, \cdots, p$, so that $\varepsilon(t)$ is uncorrelated with all the regressors in (9.9). This means that $\frac{1}{N}\sum_{t=1}^{N}\varepsilon(t)y(t-i)$ is the sample mean of N mutually uncorrelated terms with mean 0 and constant variance $E(\varepsilon(t)y(t-i))^2 < \infty$, because ε has finite fourth order moments. The weak law of large numbers implies that

$$\text{plim}(\frac{1}{N}\sum_{t=1}^{N}\varepsilon(t)y(t-i)) = 0.$$

As concerns assumption A1*, $\frac{1}{N}X_N^T X_N$ is a $p \times p$ matrix with (i, j)-th element $\frac{1}{N}\sum_{t=1}^{N}y(t-i)y(t-j)$. Under the above conditions the process y can be shown to be ergodic. The proof requires a generalized law of large numbers for the sample mean of N correlated terms (but with exponentially decaying correlation between $y(t-i)y(t-j)$ and $y(t-i+k)y(t-j+k)$ for $k \to \infty$). Ergodicity implies that the matrix Q in assumption A1* exists, and that $Q_{ij} = E(y(t-i)y(t-j))$. Further Q is invertible, because otherwise there would exist $a \in \mathbb{R}^p$ such that $a^T Q a = \text{var}(\sum_{i=1}^{p}a_i y(t-i)) = 0$ which contradicts that the autoregressive process (9.9) has no perfectly predictable component.   □

In the model (9.9) the observations have mean $Ey(t) = 0$. In practice, one may add regressors to take care of, for example, non-zero mean and trends, so that

$$y(t) = \mu_1 + \mu_2 T(t) + \alpha_1 y(t-1) + \cdots + \alpha_p y(t-p) + \varepsilon(t). \tag{9.10}$$

Least squares is also consistent for this model under the conditions of Theorem 9.4.1.

Theorem 9.4.2 *If in the autoregressive model* (9.9) *the noise ε satisfies assumption* A2* *(normality), then the least squares estimator is consistent, asymptotically efficient, and asymptotically normally distributed.*

Proof   It is sufficient to prove that under these conditions least squares is asymptotically equivalent to maximum likelihood. The likelihood function of (9.9) can be written, by conditioning, as

$$L(\alpha_1, \cdots, \alpha_p) = p(y(1), \cdots, y(N))$$
$$= p(y(1), \cdots, y(p))\Pi_{t=p+1}^{N}p(y(t) \mid y(1), \cdots, y(t-1))$$
$$= p(y(1), \cdots, y(p))\Pi_{t=p+1}^{N}p(y(t) \mid y(t-p), \cdots, y(t-1))$$
$$= p(y(1), \cdots, y(p))\Pi_{t=p+1}^{N}p(\varepsilon(t)).$$

As $p(\varepsilon(t)) = (2\pi\sigma^2)^{-\frac{1}{2}} \exp\{-(2\sigma^2)^{-1}\varepsilon(t)^2\}$ this gives

$$\frac{1}{N} \log L = \frac{1}{N} \log(p(y(1), \cdots, y(p))) + \frac{1}{N} \sum_{t=p+1}^{N} \log p(\varepsilon(t))$$

$$= \frac{1}{N} \log(p(y(1), \cdots, y(p))) - \frac{1}{2} \log(2\pi\sigma^2) - \frac{(2\sigma^2)^{-1}}{N} \sum_{t=p+1}^{N} \varepsilon(t)^2.$$

Apart from the first term, that vanishes for $N \to \infty$, this shows that the ML estimates of $\alpha_1, \cdots, \alpha_p$ are obtained by minimizing $\sum_{t=p+1}^{N} \varepsilon(t)^2$. □

There is a close connection between least squares and the so-called *Yule-Walker equations*. As $E(\varepsilon(t)y(t-i)) = 0$ for $i = 1, \cdots, p$, it follows from (9.9) that the autocovariances $R(k) = E(y(t)y(t-k))$ of the process y satisfy

$$\begin{pmatrix} R(1) \\ R(2) \\ \vdots \\ R(p) \end{pmatrix} = \begin{pmatrix} R(0) & R(1) & \cdots & R(p-1) \\ R(1) & R(0) & \cdots & R(p-2) \\ \vdots & \vdots & & \vdots \\ R(p-1) & R(p-2) & \cdots & R(0) \end{pmatrix} \begin{pmatrix} \alpha_1 \\ \alpha_2 \\ \vdots \\ \alpha_p \end{pmatrix}. \quad (9.11)$$

If we replace $R(k)$ by $\hat{R}(k) = \frac{1}{N} \sum_{t=k+1}^{N} y(t)y(t-k)$ then (9.11) can be solved for the parameters $\alpha_i, i = 1, \cdots, p$. For numerical reasons, the autocovariances are often scaled by using the correlations $\hat{\rho}(k) = \hat{R}(k)/\hat{R}(0)$ in (9.11) instead of $\hat{R}(k)$. That is, one considers estimates $\hat{\alpha}_j$ obtained by solving the following set of linear equations:

$$\begin{pmatrix} \hat{\rho}(1) \\ \hat{\rho}(2) \\ \vdots \\ \vdots \\ \hat{\rho}(p) \end{pmatrix} = \begin{pmatrix} 1 & \hat{\rho}(1) & \cdots & \cdots & \hat{\rho}(p-1) \\ \hat{\rho}(1) & 1 & \ddots & & \vdots \\ \vdots & \ddots & \ddots & \ddots & \vdots \\ \vdots & & \ddots & \ddots & \hat{\rho}(1) \\ \hat{\rho}(p-1) & \cdots & \cdots & \hat{\rho}(1) & 1 \end{pmatrix} \begin{pmatrix} \hat{\alpha}_1 \\ \hat{\alpha}_2 \\ \vdots \\ \vdots \\ \hat{\alpha}_p \end{pmatrix}. \quad (9.12)$$

The structure of the matrix in the right hand side of this equation is a very special one: it is symmetric positive definite, but also it is a Toeplitz matrix: along diagonals the same entry occurs. Fast methods to solve sets of equations of this kind for $\hat{\alpha}_1, \ldots, \hat{\alpha}_p$ are important, in particular in cases where p is large. One such fast algorithm is known as the Levinson algorithm; it requires considerably fewer numerical operations than the $O(p^3)$ operations needed for Gaussian elimination. See, e.g., [20].

To discuss the estimation of σ^2 resulting from the estimates for the α_j we use the fact that

$$\varepsilon(t) \approx \hat{\varepsilon}(t) = y(t) - \hat{\alpha}_1 y(t-1) - \cdots - \hat{\alpha}_p y(t-p).$$

Note that $\sigma^2 = E(\varepsilon(t)^2) = E(\varepsilon(t)y(t))$. Replacing in the latter formula $\varepsilon(t)$ by $\hat{\varepsilon}(t)$ we arrive at the following estimate $\hat{\sigma}^2$ for σ^2:

$$\hat{\sigma}^2 = E(\hat{\varepsilon}(t)y(t)) = \hat{R}(0) - \hat{\alpha}_1 \hat{R}(1) - \cdots - \hat{\alpha}_p \hat{R}(p).$$

One can check that the estimates resulting from solving (9.12) are approximately equal to the least squares estimates (where the summations run from $t = p+1$ to N instead of from $t = k+1$ to N).

Next we consider autoregressive models with inputs, that is,

$$y(t) = \sum_{i=1}^{p} \alpha_i y(t-i) + \sum_{i=0}^{q} \beta_i u(t-i) + \varepsilon(t) \tag{9.13}$$

Such a model is also called an ARX model, that is, an autoregressive model with exogenous variables. We assume that $\sum_{i=1}^{p} \alpha_i y(t-i) + \sum_{i=0}^{q} \beta_i u(t-i)$ is the optimal linear predictor of $y(t)$, in the sense that it minimizes the mean squared prediction error $E(y(t) - \hat{y}(t))^2$ over the class of all linear predictors of the type $\hat{y}(t) = \sum_{i \geq 0}(a_i y(t-i-1) + b_i u(t-i))$. Optimality implies that $E((y(t) - \hat{y}(t))\hat{y}(t)) = 0$, so that $E(\varepsilon(t)y(t-i)) = 0$ for all $i \geq 1$ and $E(\varepsilon(t)u(t-i)) = 0$ for all $i \geq 0$. Further we assume that the uncontrolled system with input $u(t) = 0$ is causal, that is, that $1 - \sum_{i=1}^{p} \alpha_i z^{-i}$ has all its roots inside the unit disc. We use the notation $\theta = (\alpha_1, \cdots, \alpha_p, \beta_0, \cdots, \beta_q)^T$, $x(t) = (y(t-1), \cdots, y(t-p), u(t), u(t-1), \cdots, u(t-q))^T$, and

$$Q_N = \begin{pmatrix} Q_N(yy) & Q_N(yu) \\ Q_N(uy) & Q_N(uu) \end{pmatrix} = \frac{1}{N} \sum_{t=m}^{N} x(t)x(t)^T$$

where $m = \max\{p, q\}$. So $[Q_N(yy)]_{ij} = \frac{1}{N} \sum_{t=m}^{N} y(t-i)y(t-j) = \hat{R}_y(i-j), i, j = 1, \cdots, p$, are the sample autocovariances of the output, and similarly for the other entries of the matrix Q_N.

Theorem 9.4.3 *Under the above conditions, the least squares estimators of the parameters in the ARX system (9.13) are consistent if the inputs are sufficiently excited in the sense that* $\operatorname{plim} Q_N(uu) = Q(uu)$ *exists and is invertible.*

Details of the proof fall outside the scope of this book, we refer to [21]. The idea is similar to the proof of Theorem 9.4.1. That is, the least squares estimator is given by

$\hat{\theta}_N = \theta + Q_N^{-1}\delta_N$ where $\delta_N = \frac{1}{N}\sum_{t=m+1}^{N} \varepsilon(t)x(t)$. As $\text{plim}Q_N(uu)$ exists and the system (9.13) is causal, it follows that also $\text{plim}Q_N(yy) = Q(yy)$ and $\text{plim}Q_N(yu) = Q(yu)$ exist. Further, $Q = \text{plim}Q_N$ is invertible, because otherwise there would exist $a \in \mathbb{R}^p$ and $b \in \mathbb{R}^{q+1}$ such that $(a^T, b^T)Q(a^T, b^T)^T = \text{var}(\sum_{i=1}^{p} a_i y(t-i) + \sum_{i=0}^{q} b_i u(t-i)) = 0$. Because $Q(uu)$ is invertible, $a_i \neq 0$ for at least one $i = 1, \cdots, p$, and this contradicts the fact that $y(t)$ is not perfectly predictable from the observations $\{y(s-1), u(s), s \leq t\}$. Therefore, $\text{plim}(\hat{\theta}_N) = \theta + Q^{-1}\text{plim}(\delta_N)$, and $\text{plim}(\delta_N) = 0$. This orthogonality condition again follows from a weak law of large numbers.

Note that this result does not require that the input is deterministic. It may, for instance, be generated by feedback, where $u(t)$ depends on the past outputs $\{y(s), s \leq t-1\}$. However, the input $u(t)$ may not depend on the current output $y(t)$, as in this case the orthogonality condition $E(\varepsilon(t)u(t)) = 0$ would be violated. The input condition stated in Theorem 9.4.3 can be weakened, but some persistency of excitation is needed.

In the foregoing we restricted our attention to systems (9.9) with one output and (9.13) with one input and one output. Similar results hold true for multivariate systems, with multiple inputs and outputs.

9.5 Estimation of ARMAX Models

In the foregoing section it was assumed that the disturbances $\varepsilon(t)$ in (9.9) and (9.13) are white noise. If the disturbances are correlated over time then this indicates that the dynamic specification of the model is not correct. This can be repaired by increasing the lag orders of the model, but this may lead to a large number of parameters. It may then be preferable to estimate more parsimonious models. For example, for single-input, single-output systems one can use ARMAX models defined by

$$y(t) = \sum_{i=1}^{p} \alpha_i y(t-i) + \sum_{i=0}^{q} \beta_i u(t-i) + \varepsilon(t) + \sum_{i=1}^{r} \gamma_i \varepsilon(t-i) \qquad (9.14)$$

If the inputs are $u(t) = 0$, then this is an ARMA model. We assume that this model is coprime, causal and invertible, i.e., the equations $1 - \sum_{i=1}^{p} \alpha_i z^{-i} = 0$ and $1 + \sum_{i=1}^{r} \gamma_i z^{-i} = 0$ have all their solutions in $|z| < 1$ and the equations have no common solutions. The white noise process $\varepsilon(t)$ then has the interpretation of the one-step ahead prediction errors, see Sect. 6.3.

Theorem 9.5.1 *For an ARMAX system (9.14) with $p \neq 0$ and $r \neq 0$, the least squares estimate in the regression model (9.13) is in general not consistent.*

Proof The disturbances in the model (9.13) are given by $\varepsilon(t) + \sum_{i=1}^{r} \gamma_i \varepsilon(t-i)$. If $p \neq 0 \neq r$, then these are in general correlated with the output regressors in (9.13).

Therefore the orthogonality condition is violated, and it follows from Theorem 9.2.3 that least squares is not consistent.

As a simple example, consider the ARMA(1,1) model $y(t) = \alpha y(t-1) + \varepsilon(t) + \gamma \varepsilon(t-1)$ with $\alpha \neq 0 \neq \gamma$ and $|\alpha| < 1, |\gamma| < 1$. The least squares estimate of α is given by $\hat{\alpha}_N = (\sum_{t=2}^{N} y(t) y(t-1))/(\sum_{t=2}^{N} y^2(t-1))$. From this it follows that $\mathrm{plim}(\hat{\alpha}_N) = \alpha + \gamma \sigma^2 / \mathrm{var}(y(t))$. This is inconsistent if $\gamma \neq 0$. $\square$

Consistent estimators may be obtained by using so-called *instrumental variables*. We formulate this in terms of the regression model (9.3), with $\mathrm{plim}(\frac{1}{N} X_N^T \varepsilon_N) \neq 0$ where X_N is the $N \times k$ regressor matrix and ε_N the $N \times 1$ disturbance vector for sample size N. The variables $z_i(t), i = 1, \cdots, l$, are called *instruments* if the following conditions are satisfied, where Z_N denotes the $N \times l$ matrix with elements $z_i(t)$.

$$\mathrm{plim}(\frac{1}{N} Z_N^T \varepsilon_N) = 0, \ \mathrm{plim}(\frac{1}{N} Z_N^T Z_N) = Q_{zz}, \ \mathrm{plim}(\frac{1}{N} Z_N^T X_N) = Q_{zx}$$

$$\mathrm{rank}(Q_{zz}) = l, \ \mathrm{rank}(Q_{zx}) = k. \tag{9.15}$$

The idea is to replace the regressors X_N by the instruments Z_N, because they satisfy the orthogonality condition. In order to approximate X_N as well as possible, they are regressed on Z_N. Therefore, the instrumental variables estimator $\hat{\theta}_{IV}$ is defined by the following two steps. First regress X_N on Z_N, with fitted values $\hat{X}_N = Z_N(Z_N^T Z_N)^{-1} Z_N^T X_N$, and then regress y on $\hat{X}_N$. Let $P_N = Z_N(Z_N^T Z_N)^{-1} Z_N^T$ be the projection operator on the column space of Z_N,, then

$$\hat{\theta}_{IV} = (\hat{X}_N^T \hat{X}_N)^{-1} \hat{X}_N^T y = (X_N^T P_N X_N)^{-1} X_N^T P_N y \tag{9.16}$$

Theorem 9.5.2 *The instrumental variables estimator $\hat{\theta}_{IV}$ is consistent if the conditions* (9.15) *are satisfied, and* $\mathrm{var}(\hat{\theta}_{IV})$ *is approximately given by* $\sigma^2 (X_N^T P_N X_N)^{-1}$.

Proof By filling in (9.4) into (9.16) it follows that

$$\hat{\theta}_{IV} = \theta + \{X_N^T Z_N (Z_N^T Z_N)^{-1} Z_N^T X_N\}^{-1} X_N^T Z_N (Z_N^T Z_N)^{-1} Z_N^T \varepsilon_N.$$

Consistency now follows immediately from the assumptions in (9.15). The expression for the variance follows from Theorem 9.2.2, replacing X by $\hat{X}_N$. $\square$

For the ARMAX model (9.14), assuming that the input $u(t)$ only depends on the past outputs $\{y(s), s \leq t\}$, one can choose instruments from the set $\{y(s), u(s), s \leq t - r - 1\}$ as these are uncorrelated with the composite disturbance term $\varepsilon(t) + \sum_{i=1}^{r} \gamma_i \varepsilon(t-i)$. The resulting IV estimator is consistent, but it may be far from efficient.

From an asymptotic point of view, it is optimal to use maximum likelihood. Denoting the lag operator by $(z^{-1}y)(t) = y(t-1)$, the model (9.14) can be written as

$\alpha(z^{-1})y(t) = \beta(z^{-1})u(t) + \gamma(z^{-1})\varepsilon(t)$. Because the model is assumed to be invertible, $\varepsilon(t) = (\gamma(z^{-1}))^{-1}(\alpha(z^{-1})y(t) - \beta(z^{-1})u(t)) = F(y(s), u(s), s \leq t)$ for a function F that is linear in the observed data but nonlinear in the unknown parameters $\theta = (\alpha_1, \cdots, \alpha_p, \beta_0, \cdots, \beta_q, \gamma_1, \cdots, \gamma_r)$. Because $\alpha(\infty) = \gamma(\infty) = 1$, this can also be written in prediction error form

$$\varepsilon(t, \theta) = y(t) - f(\theta, y(s-1), u(s), s \leq t) \tag{9.17}$$

If the process $\varepsilon(t)$ satisfies assumption A2*, then (conditionally on starting conditions in (9.14)) the maximum likelihood estimators are obtained by minimizing $\sum_{t=m+1}^{N} \varepsilon^2(t, \theta)$ over θ, where $m = \max\{p, q, r\}$. Note that (9.17) corresponds to a nonlinear regression model of the type (9.7), so that the parameters θ can be estimated, for instance, by the Gauss-Newton iterations (9.8).

An alternative is to use the Kalman filter. For given parameter vector θ, the ARMAX system (9.14) can be expressed in state space form, see Sect. 6.6. The mean $\mu(t)$ and variance $\sigma^2(t)$ can then be computed by means of the Kalman filter, see Theorem 7.3.1 and Proposition 7.3.3. In fact, in terms of the notation of Theorem 7.3.1 and Proposition 7.3.3 we have $\mu(t) = \hat{y}(t)$ and $\sigma^2(t) = CP(t)C^T + GG^T$. Considering the inputs as fixed and using the notation $U_t = \{u(t), u(t-1), \cdots, u(1)\}$ and similarly for Y_t, the likelihood function can be written by sequential conditioning as $\log L(\theta) = \sum_{t=1}^{N} \log(p(y(t) \mid \theta, U_t, Y_{t-1})$. Under assumption A2*, the densities $p(y(t) \mid \theta, U_t, Y_{t-1})$ are normal, with mean $\mu(t) = E(y(t) \mid \theta, U_t, Y_{t-1})$ and variance $\sigma^2(t)$, so that

$$\log L(\theta) = -\frac{N}{2}\log(2\pi) - \frac{1}{2}\sum_{t=1}^{N}(y(t) - \mu(t))^2/\sigma^2(t) - \frac{1}{2}\sum_{t=1}^{N}\log\sigma^2(t). \tag{9.18}$$

This can then serve for a numerical optimization algorithm to obtain the maximum likelihood estimate.

The foregoing results can be generalized to multivariate systems. As mentioned in Sect. 6.3, the parameters of multivariate VARMAX systems are in general not uniquely defined. That is, there exist different parameter vectors that describe exactly the same (stochastic) input-output system. This so-called non-identifiability implies that the likelihood function is constant for such parameters, so that the gradient may be zero in such directions. This causes numerical problems, that can be solved by choosing a canonical form for the parameters. We refer to [52].

Identification methods that are based on the prediction errors as in (9.17) are called prediction error identification (PEI) methods. For multivariate systems, let $V(\theta) = \frac{1}{N}\sum_{t=1}^{N}\varepsilon(t, \theta)\varepsilon^T(t, \theta)$ denote the sample covariance matrix of the prediction errors. Least squares corresponds to the criterion $\text{trace}(V(\theta))$, and it can be shown that maximum likelihood corresponds to the criterion $\log(\det(V(\theta))$. So, in the case of a single output these two methods are equivalent, but for multi-output systems this only holds true if $V(\theta)$

is diagonal and there are no cross-equation parameter restrictions in the equations (9.17). The consistency and relative efficiency of PEI methods has been investigated under quite general conditions, see [47].

9.6   Model Validation

Different model specifications may lead to different estimates of the underlying system. In order to decide about the model structure, and accordingly about the estimation method to be used, we can estimate different models and perform diagnostic tests on the underlying model assumptions. In this section we discuss some of the diagnostic tools that may be helpful in this respect.

9.6.1   Lag Orders

The estimation of ARMAX models requires that the lag orders (p, q, r) in (9.14) have been specified. If the orders are chosen too large this means that many parameters have to be estimated, with a corresponding loss of efficiency. On the other hand, if the orders are too small then the estimates become inconsistent. That is, the choice of the lag orders involves a trade-off between efficiency and consistency. We illustrate this by an example.

Example 9.6.1 Consider the causal AR(2) model $y(t) = \alpha_1 y(t-1) + \alpha_2 y(t-2) + \varepsilon(t)$, where ε satisfies assumption A2. First assume that the order is specified too large, that is, that $\alpha_2 = 0$. Using the variance expression in Theorem 9.2.2, with the regressors $x(t) = (y(t-1), y(t-2))^T$, it follows that $\hat{\alpha}_1$ in the AR(2) model has variance

$$\mathrm{var}(\hat{\alpha}_1) = \sigma^2 [(X^T X)^{-1}]_{1,1}$$

$$= \frac{\sigma^2 \sum y^2(t-2)}{\sum y^2(t-1) \sum y^2(t-2) - (\sum y(t-1)y(t-2))^2}$$

$$\approx \frac{\sigma^2}{N R(0)\{1 - (R(1)/R(0))^2\}} = \frac{1}{N},$$

where $R(k)$ denotes the autocovariances of the process $y(t)$. Because $\alpha_2 = 0$, there holds $R(0) = \sigma^2 (1 - \alpha_1^2)^{-1}$ and $R(1) = \alpha_1 R(0)$. In the correctly specified AR(1) model, the estimator has variance

$$\mathrm{var}(\hat{\alpha}_1) = \frac{\sigma^2}{\sum y^2(t-1)} \approx \frac{\sigma^2}{N R(0)} = \frac{1 - \alpha_1^2}{N}.$$

This shows that too large models lead to inefficient estimators. On the other hand, if an AR(1) model is estimated while in fact $\alpha_2 \neq 0$, then

$$\text{plim}(\hat{\alpha}_1) = \text{plim}\left(\frac{\frac{1}{N}\sum y(t)y(t-1)}{\frac{1}{N}\sum y^2(t-1)}\right) = \alpha_1 + \alpha_2 \frac{R(1)}{R(0)}. \tag{9.19}$$

So in this case the estimator is inconsistent if $R(1) \neq 0$.

Several methods have been developed for choosing the lag orders. For example, if the parameters are estimated by maximum likelihood then the results in Sect. 9.3 show that the estimators are approximately normally distributed. The significance of the parameters in model (9.14) can then be evaluated by the usual t- and F-tests.

If only a single output is observed, then the order of AR(p) models and MA(q) models can be based on the (partial) autocorrelations. The autocorrelations of a stationary process are defined by $AC(k) = R(k)/R(0)$, with corresponding sample estimates $SAC(k) = \hat{R}(k)/\hat{R}(0)$. If y is an MA(q) process, then $AC(k) = 0$ for $k > q$. If y is an AR(p) process then in the regression model (9.9) of an AR(k) model there holds $\alpha_k = 0$ for $k > p$. The sample partial autocorrelations are defined by $SPAC(k) = \hat{\alpha}_k$, the parameter of $y(t-k)$ in the estimated AR(k) model for the data (including constant, trends and dummies if needed). As a rule of thumb, estimated values SAC and $SPAC$ are considered significant if they are (in absolute value) larger than $2/\sqrt{N}$, where N is the sample size.

An alternative is to use information criteria, for instance the Akaike or Bayes criterion

$$AIC = \log(\hat{\sigma}^2) + \frac{2M}{N}, \qquad BIC = \log(\hat{\sigma}^2) + \frac{M\log(N)}{N} \tag{9.20}$$

Here $\hat{\sigma}^2$ is the estimated variance of the residuals of the model, and M is the number of AR and MA parameters of the model. For instance, for a univariate ARMA(p, q) process $M = p + q$, and for the model AR(p) model (9.10) with constant and trend $M = p$. The model with the smallest value of AIC or BIC is preferred. These criteria make an explicit trade-off between bias, measured by the error variance $\hat{\sigma}^2$, and efficiency, measured by the number of parameters.

9.6.2 Residual Tests

The estimation methods in Sects. 9.4 and 9.5 are based on the assumptions A2 or A2* for the error terms. If, for example, the lag orders have been misspecified then this may result in serial correlation of the error terms. And if the data are not appropriately transformed then the error terms may show changing variance. If the error terms are not normally distributed, then least squares is no longer equivalent to maximum likelihood. In all these cases, the methods discussed in Sects. 9.4 and 9.5 may give misleading results.

Tests of these assumptions are based on the model residuals $\hat{\varepsilon}(t) = y(t) - \hat{y}(t)$, where $\hat{y}(t)$ denotes the fitted values. For instance, for the ARMAX model (9.14) $\hat{\varepsilon}(t) = y(t) - \sum_{i=1}^{p} \hat{\alpha}_i y(t-i) - \sum_{i=0}^{q} \hat{\beta}_i u(t-i) - \sum_{i=1}^{r} \hat{\gamma}_i \hat{\varepsilon}(t-i)$. It is always informative to make a time plot of the residuals to get an idea of possible misspecification. The sample autocorrelations $SAC_\varepsilon(k) = \hat{R}_\varepsilon(k)/\hat{R}_\varepsilon(0)$ give an indication of possible serial correlation, where $\hat{R}_\varepsilon(k)$ are the sample autocovariances of $\hat{\varepsilon}(t)$. As before, if there exist many values of k for which $| SAC_\varepsilon(k) | > 2/\sqrt{N}$ then this is a sign of serial correlation.

A combined test is the Box-Pierce test $Q_m = N \sum_{k=1}^{m} (SAC_\varepsilon(k))^2$. Under the null-hypothesis that the model is correctly specified, this test follows a $\chi^2_{(m-p-r)}$ distribution for large enough sample sizes. The following Ljung-Box test involves an adjustment for finite sample effects, and also follows an asymptotic $\chi^2_{(m-p-r)}$ distribution.

$$LB_m = N(N+2) \sum_{k=1}^{m} (N-k)^{-1} (SAC_\varepsilon(k))^2. \tag{9.21}$$

The null hypothesis of no serial correlation is rejected for large values of LB_m. This means that the model is not correct, and a possible solution is to enlarge the lag orders.

As concerns heteroscedasticity, it may be that the variance is related to the level of the series or that the variance shows correlation over time. Tests are based on the series of squared residuals $\hat{\varepsilon}(t)^2$. For example, if an ARX(1, 0) model (9.13) is estimated then one can consider the regressions

$$\hat{\varepsilon}^2(t) = \lambda_0 + \lambda_1 y(t-1) + \lambda_2 y^2(t-1) + \lambda_3 u(t) + \lambda_4 u^2(t), \tag{9.22}$$

$$\hat{\varepsilon}^2(t) = \lambda_0 + \lambda_1 \hat{\varepsilon}^2(t-1) + \lambda_2 \hat{\varepsilon}^2(t-2). \tag{9.23}$$

These equations can of course be generalized. The null hypothesis is that $\lambda_i = 0$ for all $i \neq 0$. In both cases an F-test can be used, and under the null hypothesis the distribution is approximately $\chi^2_{(m)}$ where m is the number of restrictions ($m = 4$ in (9.22), and $m = 2$ in (9.23)). If there is significant heteroscedasticity then the data can be transformed, or one can adjust the identification criterion. More general, the following result holds true.

Theorem 9.6.1 *For the regression model* (9.3), *assume that A1 is satisfied and that* $E(\varepsilon) = 0$ *and* var$(\varepsilon) = V$ *with* V *nonsingular. Then the BLUE estimator is obtained by minimizing* $\varepsilon^T V^{-1} \varepsilon$, *with solution* $\hat{\beta} = (X^T V^{-1} X)^{-1} X^T V^{-1} y$ *and* var$(\hat{\beta}) = (X^T V^{-1} X)^{-1}$.

Proof As V is a nonsingular covariance matrix, it is positive definite and has a symmetric square root $V^{\frac{1}{2}}$ such that $V^{\frac{1}{2}} V^{\frac{1}{2}} = V$. Let $y_* = V^{-\frac{1}{2}} y$, $X_* = V^{-\frac{1}{2}} X$ and $\varepsilon_* = V^{-\frac{1}{2}} \varepsilon$, then (9.3) implies that $y_* = X_* \beta + \varepsilon_*$ with var$(\varepsilon_*) = I$. According to Theorem 9.2.2, the BLUE estimator is given by $\hat{\beta} = (X_*^T X_*)^{-1} X_*^T y_*$ with $\text{var}(\hat{\beta}) = (X_*^T X_*)^{-1}$, and this corresponds to the minimization of $\varepsilon_*^T \varepsilon_* = \varepsilon^T V^{-1} \varepsilon$. □

The technique to transform the data in such a way that the error term satisfies assumption A2 is called *pre-whitening*. In practice, the covariance matrix V is unknown and has to be estimated. In the case of heteroscedasticity, V is a diagonal matrix and the entries $v_{tt} = E(\varepsilon^2(t))$ can be estimated, for example, by models of the type (9.22), (9.23). The parameters β are then estimated by weighted least squares, with criterion function $\sum_{t=1}^{N} \varepsilon^2(t)/v_{tt}$.

Finally we consider the assumption of normality of the error terms. This can be tested by considering the standardized third and fourth moments of the residuals. Let $\bar{\varepsilon} = \frac{1}{N} \sum_{t=1}^{N} \hat{\varepsilon}(t)$ and $\hat{\sigma}^2 = \frac{1}{N} \sum_{t=1}^{N} (\hat{\varepsilon}(t) - \bar{\varepsilon})^2$, then $\hat{\mu}_i = \frac{1}{N} \sum_{t=1}^{N} (\hat{\varepsilon}(t) - \bar{\varepsilon})^i / \hat{\sigma}^i$ are the skewness (for $i = 3$) and kurtosis (for $i = 4$). It can be shown that, asymptotically and under the null hypothesis that A2* is satisfied, the Jarque-Bera test

$$JB = N(\frac{1}{6}\hat{\mu}_3^2 + \frac{1}{24}(\hat{\mu}_4 - 3)^2) \tag{9.24}$$

has the $\chi^2_{(2)}$ distribution. The normal distribution is symmetric (skewness zero) and has kurtosis equal to 3 (a measure of the thickness of the tails of the distribution). Normality may be rejected, for instance, because there are some excessively large residuals. They may arise because of special circumstances, for instance a measurement error or a temporary disruption of the process. Because the least squares criterion penalizes residuals by taking the squares, such outliers may have large effects on the estimates. This can be reduced by using more robust identification criteria, for example by minimizing $\sum_{t=1}^{N} | \varepsilon(t) |$.

9.6.3 Inputs and Outputs

For multivariable systems, the question arises how many equations should be estimated and what are the properties of the error process. It is usual to model either all the variables as a multivariate stochastic process or to model some of the variables (the outputs) in terms of the others (the inputs). This is also the basis for the methods described in Sects. 9.4 and 9.5. Here we will not discuss alternative modelling approaches, but we give two examples indicating the importance of these questions.

Example 9.6.2 In this example we analyse the effect of incomplete model specification. Assume that three variables are observed that actually consist of one input and two outputs, related by the equations

$$y_1(t) = \alpha_1 y_2(t) + \beta_1 y_1(t - 1) + \gamma_1 u(t) + \varepsilon_1(t),$$
$$y_2(t) = \alpha_2 y_1(t) + \beta_2 y_2(t - 1) + \gamma_2 u(t) + \varepsilon_2(t),$$

where $(\varepsilon_1, \varepsilon_2)^T$ is a white noise process with covariance matrix I. Suppose that we do not know that y_2 is an output and that we estimate only the first equation for y_1, seen as an ARX$(1, 0)$ model with output y_1 and inputs u and y_2. This model structure suggests to estimate the parameters by least squares, see Sect. 9.4. However, this gives inconsistent estimates. The result in Theorem 9.4.3 does not apply, because the regressor $y_2(t)$ is correlated with $\varepsilon_1(t)$ if $\alpha_2 \neq 0$. More precisely, assume that the processes y_1, y_2 and u are all stationary, and let $\theta = (\alpha_1, \beta_1, \gamma_1)^T$ and $x(t) = (y_2(t), y_1(t-1), u(t))^T$. Then the least squares estimator $\hat{\theta}_N$ in the equation for y_1 has the property that $\text{plim}(\hat{\theta}_N) = \theta + V^{-1}\delta$, where $V = \text{var}(x(t))$ is invertible and $\delta \in \mathbb{R}^3$ has as first entry $E(y_2(t)\varepsilon_1(t))$. Taking into account the two model equations, it follows that $E(y_2(t)\varepsilon_1(t)) = \alpha_2/(1 - \alpha_1\alpha_2) \neq 0$. This is called the simultaneity bias, that arises when some of the system equations are missing in the model.

Example 9.6.3 Next we analyse the consequences of a wrong specification of the properties of the error process. Suppose that the system consists of a single input and a single output that are both measured with error, for instance,

$$y(t) = y_*(t) + \varepsilon_1(t), \quad u(t) = u_*(t) + \varepsilon_2(t), \quad y_*(t) = \beta u_*(t - 1) + \varepsilon_3(t).$$

Here the underlying system for the unobserved variables (y_*, u_*) is ARX$(0, 1)$. We assume that ε_i are independent white noise processes with zero mean and variance σ_i^2, $i = 1, 2, 3$, and that u_* is a stationary process with mean zero and variance σ_*^2 that is independent of ε_i, $i = 1, 2, 3$. In terms of the observed input and output, the ARX$(0, 1)$ model $y(t) = \theta u(t - 1) + \varepsilon(t)$ is correctly specified, in so far as the lag order is correct, the input and output are chosen correctly, and the errors satisfy assumption A2. Indeed, actually $y(t) = \beta u(t - 1) + \varepsilon(t)$ where $\varepsilon(t) = \varepsilon_1(t) - \beta\varepsilon_2(t - 1) + \varepsilon_3(t)$ is a white noise process. However, the least squares estimator is not consistent because the orthogonality condition of Theorem 9.2.3 is not satisfied. As $E(\varepsilon(t)u(t-1)) = -\beta\sigma_2^2$ and $E(u^2(t-1)) = \sigma_*^2 + \sigma_2^2$, it follows that

$$\text{plim}(\hat{\theta}_N) = \beta - \frac{\beta\sigma_2^2}{\sigma_*^2 + \sigma_2^2} = \beta(1 - \frac{1}{S + 1}),$$

where $S = \sigma_*^2/\sigma_2^2$ is the so-called signal-to-noise ratio for the input. This shows that a wrong specification of the error assumptions may lead to inconsistent results. Especially when the noise is relatively large, that is, when S is small, the estimates may be very unreliable. Note that the orthogonality condition can not be checked by computing the correlation between the regressor $u(t - 1)$ and the residuals $\hat{\varepsilon}(t) = y(t) - \hat{\theta}u(t - 1)$, because $\text{plim}(\frac{1}{N}\sum_{t=1}^{N} \hat{\varepsilon}(t)u(t - 1)) = E(y(t)u(t - 1)) - \text{plim}(\hat{\theta}_N)E(u^2(t - 1)) = 0.$

9.6.4 Model Selection

In system identification one is confronted with the choice of data, model class, estimation method, and tools for evaluating the model quality. The validation techniques for the lag orders and the residuals discussed in Sects. 9.6.1 and 9.6.2 are of help. Further, the intended model use may suggest additional evaluation criteria. For instance, if forecasting is the objective then the models can be compared with respect to their forecast performance. The standard deviation

$$\hat{\sigma} = \{\frac{1}{N} \sum_{t=1}^{N} \hat{\varepsilon}(t)^2\}^{\frac{1}{2}} \tag{9.25}$$

is an indication of this. However, in least squares the data are first used to minimize $\hat{\sigma}$, so that this may underestimate the future forecast errors. A more reliable criterion is $\sigma^* = \{\frac{1}{N} \sum_{t=1}^{N} \varepsilon^*(t)^2\}^{\frac{1}{2}}$, where $\varepsilon^*(t) = y(t) - y^*(t)$ is the residue corresponding to the model that is estimated using the data $\{y(s-1), u(s), s \le t\}$. The disadvantage is that this requires the estimation of a sequence of models, a new one for every additional observation. One can also consider m-step-ahead prediction, where only the data $\{y(s-1), u(s), s \le t-m\}$ are used to estimate a model to forecast $y(t)$. Instead of quadratic criteria one can also consider the absolute errors $\frac{1}{N} \sum_{t=1}^{N} | \varepsilon(t) |$ or the relative errors $\frac{1}{N} \sum_{t=1}^{N} (| \varepsilon(t) | / | y(t) |)$. For input-output systems that allow experiments with the inputs, one can also compare the simulated outputs of the model with the outputs that result in reality.

Cycles and Trends

For many time series, trends and cyclical fluctuations dominate the stationary part. The main cyclical components can be identified by spectral analysis. Trends and seasonals can either be incorporated explicitly in the model or they can be removed by filtering the data.

10.1 The Periodogram

In this section we consider the modelling of a univariate time series in terms of underlying cyclical components. Let the data consist of N observations $\{y(t); t = 1, \ldots, N\}$, then a cyclical process with n components is described by

$$y(t) = \sum_{k=1}^{n} \alpha_k \sin(\omega_k t + \theta_k) + \varepsilon(t), \quad t = 1, \cdots, N. \tag{10.1}$$

The parameters θ_k are assumed to be independent and uniformly distributed on $[0, 2\pi)$. The process ε takes account of the fact that the observed time series is not purely cyclical. Here ε is assumed to be independent of the parameters θ_k, with variance $E(\varepsilon^2(t)) = \sigma^2$. The aim is to estimate the number n of cyclical components, the frequencies ω_k, and their variance contributions $\alpha_k^2, k = 1, \cdots, n$. First we assume that n and ω_k are known and that θ_k and α_k unknown. The observed time series corresponds to a single realization of the process, so that the parameters θ_k are fixed. Let $\beta_k = \alpha_k \sin\theta_k$ and $\gamma_k = \alpha_k \cos\theta_k$, then (10.1) can be written as the regression model

$$y(t) = \sum_{k=1}^{n} \beta_k \cos(\omega_k t) + \sum_{k=1}^{n} \gamma_k \sin(\omega_k t) + \varepsilon(t). \tag{10.2}$$

© The Author(s), under exclusive license to Springer Nature Switzerland AG 2021
C. Heij et al., *Introduction to Mathematical Systems Theory*,
https://doi.org/10.1007/978-3-030-59654-5_10

In the vector notation (9.3), the parameter vector is

$$q := (\beta_1, \gamma_1, \beta_2, \gamma_2, \cdots, \beta_n, \gamma_n)^T$$

and the regression matrix is

$$X = \begin{pmatrix} \cos(\omega_1) & \sin(\omega_1) & \cdots & \cos(\omega_n) & \sin(\omega_n) \\ \cos(2\omega_1) & \sin(2\omega_1) & \cdots & \cos(2\omega_n) & \sin(2\omega_n) \\ \vdots & \vdots & & \vdots & \vdots \\ \cos(N\omega_1) & \sin(N\omega_1) & \cdots & \cos(N\omega_n) & \sin(N\omega_n) \end{pmatrix},$$

that is, we rewrite (10.2) as

$$y = Xq + \varepsilon.$$

Here, $y = (y(1), \ldots, y(N))^T$, and $\varepsilon = (\varepsilon(1), \ldots, \varepsilon(N))^T$. The estimates $\hat{q} = (\hat{\beta}_1, \hat{\gamma}_1, \hat{\beta}_2, \hat{\gamma}_2, \cdots, \hat{\beta}_n, \hat{\gamma}_n)^T$ are given by

$$\hat{q} = (X^T X)^{-1} X^T y.$$

Observe also that $\tan \theta_k = \frac{\beta_k}{\gamma_k}$.

The least squares estimates of $\hat{\beta}_k, \hat{\gamma}_k$ give

$$\hat{\alpha}_k^2 = \hat{\beta}_k^2 + \hat{\gamma}_k^2 \text{ and } \hat{\theta}_k = \arctan(\hat{\beta}_k/\hat{\gamma}_k).$$

Theorem 10.1.1 *Assume that*

$$\omega_k = \frac{2\pi m_k}{N}, \quad m_k \in \mathbf{N}, \;\; 0 < m_k < \frac{N}{2}, \;\; k = 1, \cdots, n. \tag{10.3}$$

Then for the regression matrix X there holds $X^T X = \frac{N}{2} I$, and the least squares estimators in (10.2) are given by

$$\hat{\beta}_k = \frac{2}{N} \sum_{t=1}^{N} y(t) \cos(\omega_k t), \quad \hat{\gamma}_k = \frac{2}{N} \sum_{t=1}^{N} y(t) \sin(\omega_k t).$$

If in addition ε satisfies assumption A2* (Gaussian white noise), then the estimators $\hat{\alpha}_k^2$ are independent with distribution $\frac{N}{2\sigma^2} \hat{\alpha}_k^2 \sim \chi_{(2)}^2$ if $\alpha_k = 0$.

Proof To prove that $X^T X = \frac{N}{2} I$ we have to prove that

$$\sum_{t=1}^{N} \cos(\omega_j t) \sin(\omega_k t) = 0 \quad \text{for all } j, k, \tag{10.4}$$

that

$$\sum_{t=1}^{N} \cos(\omega_j t) \cos(\omega_k t) = \sum_{t=1}^{N} \sin(\omega_j t) \sin(\omega_k t) = 0 \quad \text{for all } j \neq k, \tag{10.5}$$

and that for $j = k$ the last expressions are equal to $\frac{N}{2}$. We use that

$$\cos(\omega_j t) \sin(\omega_k t) = \frac{1}{2} \sin\left((\omega_j + \omega_k)t\right) - \frac{1}{2} \sin\left((\omega_j - \omega_k)t\right),$$

$$\cos(\omega_j t) \cos(\omega_k t) = \frac{1}{2} \cos\left((\omega_j - \omega_k)t\right) + \frac{1}{2} \cos\left((\omega_j + \omega_k)t\right),$$

$$\sin(\omega_j t) \sin(\omega_k t) = \frac{1}{2} \cos\left((\omega_j - \omega_k)t\right) - \frac{1}{2} \cos\left((\omega_j + \omega_k)t\right).$$

Also for $0 < |m| < N$ we have that

$$\sum_{t=1}^{N} e^{\left(\frac{2\pi m i}{N} t\right)} = \sum_{t=1}^{N} e^{\left(\frac{2\pi m i}{N}\right)^t} = \left(e^{\left(\frac{2\pi m i}{N}(N+1)\right)} - e^{\left(\frac{2\pi m i}{N}\right)} \right) / \left(e^{\left(\frac{2\pi m i}{N}\right)} - 1 \right) = 0$$

since $\exp\left(\frac{2\pi m i}{N}(N + 1)\right) = \exp\left(\frac{2\pi m i}{N}\right)$. Then the real and imaginary parts are equal to zero and hence for $0 < |m| < N$ we have that

$$\sum_{t=1}^{N} \cos\left(\frac{2\pi m}{N} t\right) = \sum_{t=1}^{N} \sin\left(\frac{2\pi m}{N} t\right) = 0.$$

Now notice that for $j \neq k$ we have $\omega_j \pm \omega_k = \frac{2\pi}{N} m_{jk}$ with $0 < |m_{jk}| < N$, and for $j = k$ also $\omega_j + \omega_k = \frac{2\pi}{N} m_{jk}$ with $0 < |m_{jk}| < N$. Therefore for $j \neq k$

$$\sum_{t=1}^{N} \sin\left((\omega_j + \omega_k)t\right) = \sum_{t=1}^{N} \cos\left((\omega_j + \omega_k)t\right) = \sum_{t=1}^{N} \sin\left((\omega_j - \omega_k)t\right) = \sum_{t=1}^{N} \cos\left((\omega_j - \omega_k)t\right) = 0.$$

For $j = k$ we have that

$$\sum_{t=1}^{N} \sin\left((\omega_j + \omega_k)t\right) = \sum_{t=1}^{N} \cos\left((\omega_j + \omega_k)t\right) = \sum_{t=1}^{N} \sin\left((\omega_j - \omega_k)t\right) = 0,$$

and $\sum_{t=1}^{N} \cos\left((\omega_j - \omega_k)t\right) = N$. Using all this we obtain (10.4) and (10.5). This result implies the expressions given for $\hat{\beta}_k$ and $\hat{\gamma}_k$ Under assumption A2*, the covariance matrix of these estimators is $\sigma^2(X^T X)^{-1} = \frac{2\sigma^2}{N} I$, and if $\alpha_k = 0$ then $\beta_k = \gamma_k = 0$ and $\hat{\beta}_k, \hat{\gamma}_k$ are independently distributed as $N(0, \frac{2\sigma^2}{N})$. The distribution of $\hat{\alpha}_k^2$ follows by definition of the chi-square distribution.   □

So the least squares estimates of a harmonic process with frequencies satisfying (10.3) have attractive properties. The estimates of β_k and γ_k, and hence also of α_k^2 and θ_k, remain unchanged if an extra cyclical component is added to the model. The efficiency of the estimates, as measured by the variance, depends only on the number N of observations and is independent of possible misspecifications of the model like omission of a relevant frequency or inclusion of an irrelevant one. These results hold also approximately true for frequencies that do not satisfy condition (10.3).

In practice the frequencies ω_k are unknown and have to be estimated from the data. For this purpose it is helpful to analyse the observed time series in the frequency domain. A cyclical process $y(t) = \sin(\omega_t + \theta)$ has covariances $R(k) = \frac{1}{2}\cos(\omega k) = \frac{1}{4}(e^{i\omega k} + e^{-i\omega k})$. Hence, in analogy with (6.27) the spectrum is given by $S(e^{i\omega}) = \frac{1}{4}(\delta(\omega) + \delta(-\omega))$, where $\delta(\omega)$ is the Dirac distribution with the property that $\int_{-\pi}^{\pi} e^{i\lambda k}\delta(\omega)d\lambda = e^{i\omega k}$ (a point distribution with all mass at ω). So the frequency of a cyclical process is easily determined from the spectrum $S(e^{i\omega}) = \frac{1}{2\pi}\sum_{k=-\infty}^{\infty} R(k)e^{-i\omega k}$. A natural estimate of the spectrum is obtained by replacing the covariances by the sample autocovariances. This is called the periodogram.

Definition 10.1.2 The *periodogram* of the N observations $\{y(t), t = 1, \cdots, N\}$, is defined as

$$\hat{S}_N(e^{i\omega}) = \frac{1}{2\pi} \sum_{k=-(N-1)}^{N-1} \hat{R}(k)e^{-i\omega k} \tag{10.6}$$

where the autocovariances are estimated by $\hat{R}(k) := \frac{1}{N}\sum_{t=k+1}^{N} y(t)y(t-k)^T, k = 0, \cdots, N-1$ and with $\hat{R}(k) = \hat{R}(-k)^T$ for $k < 0$.

Note that on the basis of the available N observations it is not possible to compute $\hat{R}(k)$ for $k \geq N$, and in (10.6) these autocovariances are replaced by zero.

Theorem 10.1.3 *The periodogram is given by*

$$\hat{S}_N(e^{i\omega}) = \frac{1}{2\pi N} \mid \sum_{t=1}^{N} y(t)e^{-i\omega t} \mid^2 .$$

At the frequencies $\omega_k = \frac{2\pi k}{N}, 0 < k < \frac{N}{2}$, *the periodogram takes the values* $\hat{S}_N(e^{i\omega_k}) = \frac{N}{8\pi}\hat{\alpha}_k^2$ *with* $\hat{\alpha}_k^2 = \hat{\beta}_k^2 + \hat{\gamma}_k^2$ *as defined in Theorem* 10.1.1.

Proof This follows from

$$\frac{1}{N} \left| \sum_{t=1}^{N} y(t)e^{-i\omega t} \right|^2 = \frac{1}{N} \sum_{t=1}^{N}\sum_{s=1}^{N} y(t)y(s)e^{i\omega(s-t)} =$$

$$= \sum_{k=-(N-1)}^{N-1} e^{-i\omega k}\frac{1}{N} \sum_{r=|k|+1}^{N} y(r)y(r-|k|) = \sum_{k=-(N-1)}^{N-1} e^{-i\omega k}\hat{R}(k).$$

For $\omega_k = \frac{2\pi k}{N}$, and using that $\hat{\beta}_k^2 + \hat{\gamma}_k^2 = \hat{\alpha}_k^2$, we obtain from Theorem 10.1.1 that

$$\hat{S}_N(e^{i\omega_k}) = \frac{1}{2\pi N} \left(\{\sum_{t=1}^{N} y(t)\cos(\omega_k t)\}^2 + \{\sum_{t=1}^{N} y(t)\sin(\omega_k t)\}^2 \right) = \frac{N}{8\pi}\hat{\alpha}_k^2.$$

□

Theorem 10.1.4 *Let y be a cyclical process* (10.1) *where ε is Gaussian white noise and with frequencies* ω_k *satisfying condition* (10.3). *The periodogram then has the properties that*

$$E\{\hat{S}_N(e^{i\omega_k})\} = \frac{1}{8\pi}(N\alpha_k^2 + 4\sigma^2), \tag{10.7}$$

$$\frac{4\pi}{\sigma^2}\hat{S}_N(e^{i\omega_k}) \sim \chi_{(2)}^2 \text{ if } \alpha_k = 0. \tag{10.8}$$

Proof The result in (10.8) follows from Theorems 10.1.1 and 10.1.3. It further follows from Theorem 9.2.2, with $(X^T X)^{-1} = \frac{2}{N}I$ according to Theorem 10.1.1, that $E\{\hat{S}_N(e^{i\omega_k})\} = \frac{N}{8\pi}E(\hat{\beta}_k^2 + \hat{\gamma}_k^2) = \frac{N}{8\pi}\{\text{var}(\hat{\beta}_k) + \text{var}(\hat{\gamma}_k) + (E\hat{\beta}_k)^2 + (E\hat{\gamma}_k)^2\} = \frac{1}{8\pi}(4\sigma^2 + N\alpha_k^2)$.

□

This means that the periodogram tends linearly to infinity at the frequencies that are present in the process. If the cycle with frequency $\frac{\omega_k}{2\pi}$ is absent, then the periodogram has a finite average value. So, on average, the periodogram can clearly detect the cyclical components. *If the periodogram shows a peak around a certain frequency, then this indicates that cycles with this frequency contribute substantially to the variations in the process.*

It should be mentioned that the sampling period may influence the location of the peaks in the periodogram. As a simple example, if the process $y(t) = \sin(\omega t)$ is observed at time instants $t = k\Delta$, $k \in \mathbf{N}$, then the frequencies $\omega + \frac{2\pi l}{\Delta}$, $l \in \mathbf{N}$, can not be discriminated by the data. This effect is called *aliasing*.

Example 10.1.1 As an illustration, we consider the cyclical process (10.1) with $n = 2$ frequencies. Here ε is a white noise process with variance $\sigma^2 = 1$, θ_1 and θ_2 are randomly chosen on $[0, 2\pi)$, $\omega_k = \frac{\pi k}{10}$ and $\alpha_k = k$, $k = 1, 2$. The periodograms for sample sizes $N = 16$, $N = 128$ and $N = 1024$ are shown in the following figure.

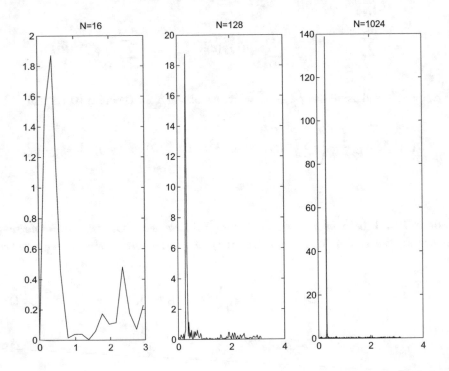

The computation of the periodogram, when done via direct calculations as described in Definition 10.1.2 for all frequencies $\omega_k = \frac{2\pi k}{N}$, $k = 0, \cdots, N - 1$, requires the order of N^2 operations of multiplication and addition. This can be reduced to the order $N \log(N)$ by the Fast Fourier transform (FFT). The idea is as follows.

Proposition 10.1.5 *Let* $N = n_1 n_2$ *and consider the periodogram at the frequencies* $\omega_k = \frac{2\pi k}{N}, k = 0, \cdots, N-1$, *where* $k = n_2 k_1 + k_2, k_1 = 0, \cdots, n_1 - 1, k_2 = 1, \cdots, n_2$. *Then*

$$\hat{S}_N(e^{i\omega_k}) = \frac{1}{2\pi N} \mid \sum_{t_1=1}^{n_1} e^{\frac{2\pi i k t_1}{N}} \{\sum_{t_2=0}^{n_2-1} y(n_1 t_2 + t_1) e^{\frac{2\pi i k_2 t_2}{n_2}}\} \mid^2 . \tag{10.9}$$

The number of operations involved in calculating (10.7) for all $k = 0, \cdots, N-1$ *is of the order* $N(n_1 + n_2)$.

Proof Let $t = n_1 t_2 + t_1$, then (10.9) follows directly from the periodogram formula in Theorem 10.1.3, using the fact that

$$\exp\left(\frac{2\pi i k t}{N}\right) = \exp\left(\frac{2\pi i k t_1}{N}\right) \exp\left(\frac{2\pi i (n_2 k_1 + k_2) n_1 t_2}{N}\right)$$

$$= \exp\left(\frac{2\pi i k t_1}{N}\right) \exp\left(\frac{2\pi i k_2 t_2}{n_2}\right).$$

Now observe that the term in braces in (10.9) depends on k only via k_2, not via k_1. For fixed t_1 and k_2 this term requires the order of n_2 additions and multiplications. In total this gives $n_1 n_2^2$ operations for all possible terms in brackets. The term in front of it implies n_1 operations for each value of k, so in total this needs an additional number of $N n_1$ operations. The total number is hence $N n_1 + n_1 n_2^2 = N(n_1 + n_2)$. $\square$

By a similar factorization $N = \Pi_{i=1}^k n_i$ this leads to the order of $N(\sum_{i=1}^k n_i)$ operations, and if $N = 2^n$ this gives $2nN$ which is of the order $N \log(N)$.

10.2 Spectral Identification

In most observed time series there are no sharply defined frequencies that dominate the fluctuations. Instead there may exist certain bands of frequencies that are relatively more important than other ones. Instead of the cyclical process (10.1) that has discrete spectrum, such series are better described by the continuous spectrum of a moving average process. In this section we consider frequency domain identification methods for univariate stationary processes and for single input, single output systems.

First we consider the case that the observed process is white noise. The periodogram is not a completely satisfactory estimate of the spectrum in this case.

Theorem 10.2.1 *The periodogram of a white noise process with variance σ^2 is an unbiased but inconsistent estimate of the theoretical spectrum $S(e^{i\omega}) = \frac{\sigma^2}{2\pi}$, that is, for $\omega_k = \frac{2\pi k}{N}, 0 < k < \frac{N}{2}$, there holds*

$$E\{\hat{S}_N(e^{i\omega_k})\} = \frac{\sigma^2}{2\pi} = S(e^{i\omega_k}), \tag{10.10}$$

$$\text{var}\{\hat{S}_N(e^{i\omega_k})\} = \frac{\sigma^4}{4\pi^2} = S^2(e^{i\omega_k}). \tag{10.11}$$

Furthermore, the periodogram estimates $\hat{S}_N(e^{i\omega_k})$ and $\hat{S}_N(e^{i\omega_l})$ are uncorrelated for $k \neq l$.

Proof   A white noise process satisfies equation (10.1) with all coefficients $\alpha_k = 0$. So (10.10) follows from (10.7) and (10.11) from (10.8) because a $\chi^2_{(2)}$ distribution has variance equal to 4. That the estimates are uncorrelated follows from Theorems 10.1.1 and 10.1.3.   □

This shows that the periodogram is an unbiased estimate of the spectrum but that the variance does not decrease when the sample size increases. Moreover, as adjoining estimates are uncorrelated the periodogram typically has a very irregular shape. Similar results hold true for other processes with continuous spectrum.

Theorem 10.2.2 *If the data are generated by a moving average process $y(t) = \sum_{k=0}^{\infty} G_k \varepsilon(t - k)$ with $\sum_{k=0}^{\infty} G_k^2 < \infty$, then the periodogram at frequencies $\omega_k = \frac{2\pi k}{N}$ has the properties that*

$$\lim_{N \to \infty} E\{\hat{S}_N(e^{i\omega_k})\} = S(e^{i\omega_k}),$$

$$\lim_{N \to \infty} \text{var}\{\hat{S}_N(e^{i\omega_k})\} = S^2(e^{i\omega_k}),$$

$$\lim_{N \to \infty} \text{cov}\{\hat{S}_N(e^{i\omega_k}), \hat{S}_N(e^{i\omega_l})\} = 0 \qquad \text{for } k \neq l.$$

Proof   We will not give a complete proof. Instead we will make the results plausible. According to Theorem 10.1.3 the periodogram is given by

$$\hat{S}_N(e^{i\omega} = \frac{1}{2\pi N} \left| \sum_{t=1}^{N} y(t) e^{-i\omega t} \right|^2.$$

Since $y(t) = \sum_{k=0}^{\infty} G_k \varepsilon(t-k)$, it follows that

$$\hat{S}_N(e^{i\omega}) = \frac{1}{2\pi N}\left|\sum_{t=1}^{N}\sum_{k=0}^{\infty} G_k \varepsilon(t-k)e^{-i\omega t}\right|^2 = \frac{1}{2\pi N}\left|\sum_{k=0}^{\infty} G_k e^{-i\omega k}\sum_{t=1}^{N}\varepsilon(t-k)e^{-i\omega(t-k)}\right|^2.$$

Next note that according to Theorem 10.1.3 we also have

$$\hat{S}_N^{\varepsilon}(e^{i\omega}) = \frac{1}{2\pi N}\left|\sum_{t=1}^{N}\varepsilon(t)e^{-i\omega t}\right|^2.$$

Then

$$2\pi S(e^{i\omega})\hat{S}_N^{\varepsilon}(e^{i\omega})) = \left|G(e^{-i\omega})\right|^2 \hat{S}_N^{\varepsilon}(e^{i\omega}) = \frac{1}{2\pi N}\left|\sum_{k=0}^{\infty} G_k e^{-i\omega k}\sum_{t=1}^{N}\varepsilon(t)e^{-i\omega t}\right|^2$$

This way it is plausible that

$$\lim_{N\to\infty}\hat{S}_N(e^{i\omega}) - \left|G(e^{i\omega})\right|^2 \hat{S}_N^{\varepsilon}(e^{i\omega}) = 0 \tag{10.12}$$

Also we have that $S(e^{i\omega}) = \frac{1}{2\pi}G(e^{i\omega})G(e^{-i\omega})$. According to Theorem 10.2.1, $\hat{S}_N^{\varepsilon}$ has mean $\frac{1}{2\pi}$ and variance $\frac{1}{4\pi^2}$, and the values at different frequencies are uncorrelated. Using this in (10.12) the results follow.   □

If the spectrum is continuous then *adjoining estimates can be used for smoothing to reduce the variance*. However, this may introduce a bias as sharp peaks in the spectrum are smoothed over a larger region, the so-called leakage effect. Various consistent smoothing procedures have been developed that differ in their bias and variance properties. The idea is as follows. For given frequency ω and $M < N$, let I be the set of M integers k for which $\omega_k = \frac{2\pi k}{N}$ are closest to ω. Then estimate the spectrum by the average $S_N^*(e^{i\omega}) = \frac{1}{M}\sum_{k\in I}\hat{S}_N(e^{i\omega_k})$. This introduces a bias, but according to Theorem 10.2.2, $E(S_N^*(e^{i\omega})) = \frac{1}{M}\sum_{K\in I} S(e^{i\omega_k})$, and if S is continuous this converges to $S(e^{i\omega})$ if $\frac{M}{N} \to 0$. Because the periodogram estimates are uncorrelated, it follows that $\mathrm{var}(S_N^*(e^{i\omega})) = \frac{1}{M^2}\sum_{k\in I} S^2(e^{i\omega_k}) \to 0$ if $M \to \infty$. This shows that consistent estimates are obtained by smoothing, provided that the smoothing interval $\frac{M}{N} \to 0$ and the number of included frequencies $M \to \infty$. Other smoothed estimates are obtained by

$$S_N^*(e^{i\omega}) = \int_{-\pi}^{\pi} F(\omega - \lambda)\hat{S}_N(e^{i\lambda})d\lambda$$

where F is the smoothing filter. The foregoing corresponds to a uniform filter and is called the Daniell window. In practice one often uses filters that are more smooth, for instance, Hamming and Bartlett windows.

Example 10.2.1 The spectrum of an AR(1) process $y(t) = \alpha y(t-1) + \varepsilon(t)$ is given by $\sigma^2 \{2\pi(1 + \alpha^2 - 2\alpha\cos\omega)\}^{-1}$, see Example 6.5.1. The reader is urged to produce (for instance using Matlab) graphs of the spectra and (smoothed) periodograms for the cases $\alpha = 0$ and $\alpha = 0.9$, for sample sizes $N = 128$ and $N = 1024$. The smoothing can be done for instance as follows: replace in the formula for the periodogram the $\hat{R}(k)$'s by $\hat{S}(k) = (1 - \frac{|k|}{M})R(k)$, for $k = -M + 1, \cdots, M - 1$, and $\hat{S}(k) = 0$ for $|k| \geq M$, where M is (much) smaller than N. For example, for the sample sizes indicated one may take $M = 15$.

Clearly, smoothing reduces the variance and the bias diminishes for larger sample sizes. Because the spectrum of a white noise process is constant, smoothing introduces no bias when $\alpha = 0$.

Smoothed periodograms can be interpreted as nonparametric estimates of stationary processes. This method of spectral identification can be extended to input-output systems. For example, consider the single input, single output system described by

$$y(t) = \sum_{k=0}^{\infty} G_k u(t-k) + \varepsilon(t). \tag{10.13}$$

Theorem 10.2.3 *The least squares estimate of the transfer function $G(z)$ is approximately given by*

$$\hat{G}(e^{i\omega}) = \hat{S}_{yu}(e^{i\omega})/\hat{S}_{uu}(e^{i\omega}) \tag{10.14}$$

where $\hat{S}_{yu}(e^{i\omega}) = \frac{1}{2\pi}\sum_{k=-(N-1)}^{N-1} \hat{R}_{yu}(k)e^{-i\omega k}$ with

$$\hat{R}_{yu}(k) = \frac{1}{N}\sum_{t=|k|+1}^{N} y(t)u(t-k)$$

and where $\hat{S}_{uu} = \frac{1}{2\pi}\sum_{k=-(N-1)}^{N-1} \hat{R}_{uu}(k)e^{-i\omega k}$ with

$$\hat{R}_{uu}(k) = \frac{1}{N}\sum_{t=|k|+1}^{N} u(t)u(t-k)$$

With a small calculation we shall make the result plausible. First, we neglect the fact that the summations are finite. The least squares criterion corresponds to finding G_k's which minimize

$$\sum_t \| y(t) - \sum_l G_l u(t-l) \|^2.$$

The first order conditions for a minimum imply that

$$\sum_t \{ y(t) - \sum_l \hat{G}_l u(t-l) \} u(t-k) = 0,$$

so that $\hat{R}_{yu}(k) = \sum_l \hat{G}_l \hat{R}_{uu}(k-l)$. Therefore

$$\hat{S}_{yu}(e^{i\omega}) = \frac{1}{2\pi} \sum_k \sum_l \hat{G}_l \hat{R}_{uu}(k-l) e^{-i\omega k} =$$

$$= \sum_l \hat{G}_l e^{-i\omega l} \frac{1}{2\pi} \sum_k \hat{R}_{uu}(k-l) e^{-i\omega(k-l)} \approx \hat{G}(e^{i\omega}) \hat{S}_{uu}(e^{i\omega}).$$

The approximation error tends to zero for $N \to \infty$.

This is called the empirical transfer function estimate. It can be expressed in terms of the so-called discrete Fourier transform of the observations, defined as

$$y(e^{i\omega}) := \frac{1}{\sqrt{N}} \sum_{t=1}^{N} y(t) e^{-it\omega}, \quad 0 \le \omega \le \pi. \tag{10.15}$$

Theorem 10.2.4 *The approximate least squares estimate in* (10.14) *is equal to*

$$\hat{G}(e^{i\omega}) = y(e^{i\omega})/u(e^{i\omega}). \tag{10.16}$$

Proof   Recall that the we took $u(t) = 0$ and $y(t) = 0$ whenever $t < 1$ or $t > N$. Then

$$\hat{S}_{yu}(e^{i\omega}) = \frac{1}{2\pi N} \sum_{k=-(N-1)}^{N-1} \sum_{t=1}^{N} y(t) u(t-k) e^{-i\omega k}$$

$$= \frac{1}{2\pi N} \sum_{k=-(N-1)}^{N-1} \sum_{t=1}^{N} y(t) e^{-i\omega t} u(t-k) e^{i\omega(t-k)}$$

$$= \frac{1}{2\pi N} \sum_{t=1}^{N} y(t) e^{-i\omega t} \sum_{k=-(N-1)}^{N-1} u(t-k) e^{i\omega(t-k)}$$

$$= \frac{1}{2\pi N} \sum_{t=1}^{N} y(t) e^{-i\omega t} \sum_{s=1}^{N} u(s) e^{i\omega s} = \frac{1}{2\pi} y(e^{i\omega}) u^*(e^{i\omega}).$$

Replacing y by u in the above computation gives $\hat{S}_{uu}(e^{i\omega}) = \frac{1}{2\pi}u(e^{i\omega})u^*(e^{i\omega})$. The result follows from (10.14). $\square$

According to Chap. 2, in the noise-free case the relation between input and output is described in the frequency domain by $y(e^{i\omega}) = G(e^{i\omega})u(e^{i\omega})$. This provides a direct motivation for (10.16). To investigate the asymptotic properties of this estimation method, suppose that the input and output are related by the system (10.13) with $\sum_{k=0}^{\infty} G_k^2 < \infty$ and with u and ε independent. We assume that the input is stationary in the sense that $S_{uu}(e^{i\omega}) = \lim_{N\to\infty} \hat{S}_{uu}(e^{i\omega})$ exists and that ε is stationary with zero mean and with spectrum $S_{\varepsilon\varepsilon}$ (it need not be white noise).

Theorem 10.2.5 *Under the above conditions, and with $\sum_{k=-\infty}^{\infty} |kR_{\varepsilon\varepsilon}(k)| < \infty$, the least squares estimator (10.13) of the transfer function has the following properties:*

$$\lim_{N\to\infty} E\{\hat{G}(e^{i\omega})\} = G(e^{i\omega}), \tag{10.17}$$

$$\lim_{N\to\infty} \mathrm{var}\{\hat{G}(e^{i\omega})\} = S_{\varepsilon\varepsilon}(e^{i\omega})/S_{uu}(e^{i\omega}), \tag{10.18}$$

$$\lim_{N\to\infty} \mathrm{cov}\{\hat{G}(e^{i\omega_1}), \hat{G}(e^{i\omega_2})\} = 0 \text{ for } \omega_1 \neq \omega_2. \tag{10.19}$$

Sketch of a Proof We only give the main ideas. From (10.13) we obtain that, for N sufficiently large, $y(e^{i\omega}) \approx G(e^{-i\omega})u(e^{i\omega}) + \varepsilon(e^{i\omega})$. So (10.16) shows that $\hat{G}(e^{-i\omega}) \approx G(e^{-i\omega}) + \varepsilon(e^{i\omega})/u(e^{i\omega})$. Then (10.17) is evident, as u is independent of ε and $E\{\varepsilon(e^{i\omega})\} = 0$ for all ω. The variance is given by

$$E\{\varepsilon(e^{i\omega})/u(e^{i\omega})\}\{\varepsilon(e^{-i\omega})/u(e^{-i\omega})\} \approx S_{\varepsilon\varepsilon}(e^{i\omega})/S_{uu}(e^{i\omega}).$$

For example,

$$E\{\varepsilon(e^{i\omega})\varepsilon(e^{-i\omega})\} = E\{\frac{1}{N}\sum_{t=1}^{N}\sum_{s=1}^{N} \varepsilon(t)\varepsilon(s)e^{-i(t-s)\omega}\}$$

$$= E\{\sum_{k=-(N-1)}^{N-1} e^{-ik\omega}\frac{1}{N}\sum_{t=k+1}^{N} \varepsilon(t)\varepsilon(t-k)\} \approx$$

$$\approx \sum_{k=-(N-1)}^{N-1} e^{-ik\omega} R_{\varepsilon\varepsilon}(k) \approx 2\pi S_{\varepsilon\varepsilon}(e^{i\omega}).$$

To prove (10.19), the foregoing shows that $\hat{G}(e^{i\omega}) - E(\hat{G}(e^{i\omega})) \approx \varepsilon(e^{i\omega})/u(e^{i\omega})$, so it suffices to prove that, for $k \neq l$ and N sufficiently large, $E(\varepsilon(e^{i\omega_k})\varepsilon^*(e^{i\omega_l})) \approx 0$. Now

$$E(\varepsilon(e^{i\omega_k})\varepsilon^*(e^{i\omega_l})) = \frac{1}{N}\sum_{r=1}^{N}\sum_{s=1}^{N} e^{i(\omega_l s - \omega_k r)} R_{\varepsilon}(r-s) =$$

$$= \frac{1}{N}\sum_{r=1}^{N} e^{i(\omega_l - \omega_k)r} \sum_{t=r-N}^{r-1} R_{\varepsilon}(t)e^{-i\omega_l t} =$$

$$= S_{\varepsilon\varepsilon}(e^{i\omega_l})\frac{1}{N}\sum_{r=1}^{N} e^{i(\omega_l - \omega_k)r} - a_N,$$

where $a_N = \frac{1}{N}\sum_{r=1}^{N} e^{i(\omega_l - \omega_k)r} \{\sum_{t=-\infty}^{r-N-1} R_{\varepsilon}(t)e^{-i\omega_l t} + \sum_{t=r}^{\infty} R_{\varepsilon}(t)e^{-i\omega_l t}\}$. Using the fact that $R_{\varepsilon}(-t) = R_{\varepsilon}(t)$ for $t > 0$ and collecting terms, one sees $|a_N| \leq \frac{2}{N}\sum_{k=-\infty}^{\infty} |kR_{\varepsilon}(k)|$ and therefore $a_N \to 0$ for $N \to \infty$.

Finally, for $k \neq l$ there holds $\omega = \omega_l - \omega_k = \frac{2\pi m}{N}$ for some $m \in \{\pm 1, \pm 2, \cdots, \pm(N-1)\}$, so that $\sum_{r=1}^{N} e^{i(\omega_l - \omega_k)r} = \sum_{r=1}^{N} e^{i\omega r} = (1 - e^{i\omega})^{-1}(e^{i\omega} - e^{i\omega(N+1)}) = (1 - e^{i\omega})^{-1}e^{i\omega}(1 - e^{2\pi im}) = 0$. This completes the sketch of the proof of Theorem 10.2.5.

The expression (10.18) shows that the estimator has smaller variance for frequencies which are relatively strongly present in the input signal. The variance is inversely proportional to the signal-to-noise ratio. Consistency is obtained after smoothing the periodograms $\hat{S}_{yu}$ and $\hat{S}_{uu}$ in (10.14), provided that the input is sufficiently exciting in the sense that $S_{uu}(e^{i\omega}) > 0$ for all frequencies ω.

In the foregoing we considered least squares estimation of the unrestricted model (10.13). If the model is expressed in terms of finitely many parameters, then these parameters can also be estimated in the frequency domain. As an example we consider the ARMAX model (9.14), which in terms of the lag polynomials $\alpha(z) = 1 - \sum_{i=1}^{p} \alpha_i z^{-i}$, $\beta(z) = \sum_{i=0}^{q} \beta_i z^{-i}$ and $\gamma(z) = 1 + \sum_{i=1}^{r} \gamma_i z^{-i}$ can be written as

$$\alpha(e^{i\omega})y(e^{i\omega}) = \beta(e^{i\omega})u(e^{i\omega}) + \gamma(e^{i\omega})\varepsilon(e^{i\omega}). \tag{10.20}$$

Assume that the model is stationary and invertible, as defined in Sect. 6.3, and that ε is Gaussian white noise.

Theorem 10.2.6 *The maximum likelihood estimators of the ARMAX model (10.20) are obtained, for $N \to \infty$, as the minimum of*

$$\int_{-\pi}^{\pi} \left| \frac{y(e^{i\omega})}{u(e^{i\omega})} - \frac{\beta(e^{i\omega})}{\alpha(e^{i\omega})} \right|^2 \cdot \left| \frac{\alpha(e^{i\omega})}{\gamma(e^{i\omega})} \right|^2 \cdot S_{uu}(e^{i\omega})d\omega. \tag{10.21}$$

Proof Because ε is Gaussian white noise, ML corresponds to the minimization of $\frac{1}{N}\sum_{t=1}^{N}\varepsilon^2(t) = \hat{R}_{\varepsilon\varepsilon}(0) = \int_{-\pi}^{\pi}\hat{S}_{\varepsilon\varepsilon}(e^{i\omega})d\omega = \frac{1}{2\pi}\int_{-\pi}^{\pi}|\hat{\varepsilon}(\omega)|^2 d\omega$, where $\hat{\varepsilon}(e^{i\omega})$ is the discrete Fourier transform defined in (10.15). It follows from the time domain equation (10.20), with z the lag operator, that (for N sufficiently large) $\gamma(e^{i\omega})\hat{\varepsilon}(e^{i\omega}) \approx \alpha(e^{i\omega})y(e^{i\omega}) - \beta(e^{i\omega})u(e^{i\omega})$. Therefore $\hat{\varepsilon}(e^{i\omega}) \approx \frac{\alpha(e^{i\omega})}{\gamma(e^{i\omega})}u(e^{i\omega})\{\frac{y(e^{i\omega})}{u(e^{i\omega})} - \frac{\beta(e^{i\omega})}{\alpha(e^{i\omega})}\}$. This shows (10.21).

$\square$

The interpretation is that the nonparametric estimator (10.16) is approximated by the transfer function β/α of the ARMAX model (10.20). Each frequency has a weighting factor, determined by the inverse noise filter α/γ and the input spectrum S_{uu}. The approximation by the parametric model (10.20) will be most accurate where the noise filter γ/α has smallest amplitude, as for these frequencies $|\alpha(e^{i\omega})/\gamma(e^{i\omega})|$ is relatively large so that errors at these frequencies are heavily penalized. The approximation in certain frequency regions can be improved by giving the inputs relatively larger power $|u(e^{i\omega})|$ for such frequencies. This will be at the expense of worse approximations at other frequencies.

10.3 Trends

From here on the chapter will have more an overview character, and many details will be omitted. For proofs and details we refer to the literature on time series, e.g., [10].

For many time series the trending pattern is the most dominant characteristic. For purposes of forecasting and control, it is crucial to take appropriate account of the trend. The two main approaches are data transformation and explicit trend modelling. In the first case the data are transformed to obtain stationarity, and the identified model for the stationary data can be transformed into a model for the original data. In the second case the model contains, apart from a stationary part, also variables that model the trend explicitly.

In order to remove the trend by transforming the data we need a model for the trend component. For instance, if an economic variable is expressed in nominal terms then it often shows exponential growth over time because of price inflation. This can be removed by expressing the variable in real terms, by dividing it by a price index series. Also in real terms, many series still exhibit exponential growth. By taking the logarithm this transforms into a more linear trend pattern. In this case the trend can be estimated as the local average of the series. This is called smoothing. If y denotes the observed series and T the trend, then $T(t) = \sum_{k=-\infty}^{\infty}\beta_k y(t-k)$ with $\beta_k \geq 0$ and $\sum_{k=-\infty}^{\infty}\beta_k = 1$. The current trend can only be estimated if the filter is causal, that is, if $\beta_k = 0$ for all $k < 0$. A popular method is exponential smoothing with coefficients $\beta_k = \beta(1-\beta)^k$, $k \geq 0$, for some $0 < \beta < 1$. This assigns a larger weight to more recent observations, and the trend can be expressed recursively as

$$T(t) = T(t-1) + \beta(y(t) - T(t-1)). \tag{10.22}$$

The parameter β is called the forgetting factor. If β is small this produces smooth trends, and if β is nearly one then the trend follows the fluctuations in the process very rapidly. If the time series shows a relatively stable trend, that is, if $y(t) - T(t-1)$ is more or less constant over time, then (10.22) can be written as the deterministic linear trend

$$T(t) = \mu_1 + \mu_2 t. \tag{10.23}$$

This can be extended to other time functions, for instance a quadratic trend $T(t) = \mu_1 + \mu_2 t + \mu_3 t^2$ or a trend with saturation $T(t) = \mu_1(1 + \mu_2 e^{-\mu_3 t})^{-1}$. The parameters μ_i of these time functions can be estimated, for instance by least squares, replacing $T(t)$ by the observed series $y(t)$.

Another type of trend is expressed by stochastic models, for instance the random walk with drift

$$y(t) = \mu + y(t-1) + \varepsilon(t), \tag{10.24}$$

where ε is a stationary process. For such processes the trend can be removed by taking the first difference $\Delta y(t) = y(t) - y(t-1)$. The process y in (10.24) is called integrated of order 1. More general, y is an ARIMA(p, d, q) process if $\Delta^d y$ is a stationary ARMA(p, q) process, whereas $\Delta^{d-1} y$ is non-stationary. This process can therefore be described as

$$\alpha(z)(1 - z^{-1})^d y = \beta(z)\varepsilon \tag{10.25}$$

with ε white noise and where $\alpha(z)$ and $\beta(z)$ have all their roots inside the unit disc. To estimate an ARIMA model, we can follow the procedures described in Chap. 9 once the order of integration d has been selected. In practice, often $d = 0$ (so that there are no trends) or $d = 1$. A simple method to test whether $d = 0$ or $d = 1$ is to consider the autocorrelations $\rho(k) = R(k)/R(0)$ of the process. If the process is a stationary ARMA process then the autocorrelations tend exponentially to zero for $k \to \infty$. On the other hand, if y is a random walk process (10.24) without drift ($\mu = 0$) and with $y(0) = 0$ then $\rho(t, k) = [Ey(t)y(t-k)][Ey^2(t)Ey^2(t-k)]^{-\frac{1}{2}} = (1 - \frac{k}{t})^{\frac{1}{2}}$. The sample autocorrelations will decrease only slowly, as for small k there holds $(1 - \frac{k}{t})^{\frac{1}{2}} \approx 1 - \frac{k}{t}$ which gives a linear decline instead of an exponential one. The order of integration d can be chosen such that $\Delta^{d-1} y$ has a linear decline but $\Delta^d y$ has an exponential decline of the autocorrelations. Alternatively, one can also test for the unit coefficient of $y(t-1)$ in (10.24) against the stationary alternative that $y(t) = \mu + \alpha y(t-1) + \varepsilon(t)$ with $|\alpha| < 1$. By defining $\rho = \alpha - 1$, this can be written as

$$\Delta y(t) = \mu + \rho y(t-1) + \varepsilon(t), \qquad H_0 : \rho = 0. \tag{10.26}$$

The so-called Dickey-Fuller test is the t-statistic of ρ obtained from the regression in (10.26). If ε is not a white noise process, the model can be extended, for example, to

$$\Delta y(t) = \mu_1 + \mu_2 t + \rho y(t-1) + \sum_{i=1}^{k} \gamma_i \Delta y_{t-i} + \varepsilon(t), \qquad H_0 : \rho = 0. \qquad (10.27)$$

The null hypothesis of a stochastic trend ($\rho = 0$) is rejected for values of $\rho = \alpha - 1$ that are significantly smaller than zero. Because the regressor $y(t-1)$ in (10.27) is not stationary under the null hypothesis, the standard regression theory does not apply in this case. This is because assumption A1* in Chap. 9 is not satisfied, as $\text{plim}(\frac{1}{N}\sum_{t=2}^{N} y^2(t-1)) = \infty$. Critical values for the t-statistic of ρ can therefore not be obtained from the t-distribution. For example, at 5% significance level the critical value (for $N \to \infty$) of the t-distribution is -1.65, whereas for the test in (10.27) it is -3.41.

In models like (10.22), (10.23) and (10.24) the trend is modelled directly in terms of the observations. An alternative is a model with latent trend variable, for example

$$T(t+1) = \varphi T(t) + \varepsilon_1(t), \qquad y(t) = T(t) + \varepsilon_2(t). \qquad (10.28)$$

Assume that $\varepsilon = (\varepsilon_1, \varepsilon_2)^T$ is a Gaussian white noise process with mean zero and covariance matrix $\begin{pmatrix} \sigma_1^2 & 0 \\ 0 & \sigma_2^2 \end{pmatrix}$. If $|\varphi| < 1$ then y is a stationary process, if $\varphi = 1$ then y is integrated of order $d = 1$, and if $\varphi > 1$ then y grows exponentially. This is a stochastic state space model of the form (7.8), (7.9), with parameters $A = \varphi$, $B = 0$, $C = 1$, $D = 0$, $F = (\sigma_1, 0)$ and $G = (0, \sigma_2)$. The trend acts as a state variable that can be estimated by the Kalman filter, see Theorem 7.3.1. Let $\hat{T}(t+1) = E(T(t+1)|y(s), 0 \le s \le t)$, then

$$\hat{T}(t+1) = \varphi \hat{T}(t) + k(t)(y(t) - \hat{T}(t)). \qquad (10.26)$$

This is of the form (10.19) with forgetting factor $k(t)$ computed by the Kalman filter equations (7.13), (7.14). In particular, for $\varphi = 1$ the process y has a stochastic trend. In Example 9.6.3 it was shown that, for $t \to \infty$, the forgetting factor k is small if the signal-to-noise ratio σ_1^2/σ_2^2 is small, and k is nearly one if this ratio is large.

10.4   Seasonality and Nonlinearities

Seasonal variation may occur for time series that are observed, for example, every quarter or every month. For ease of exposition we will assume that the data consists of quarterly observations, but the following can be generalized to other observation frequencies.

A deterministic model is given by

$$S(t) = \mu_1 D_1(t) + \mu_2 D_2(t) + \mu_3 D_3(t) + \mu_4 D_4(t), \tag{10.29}$$

where $D_i(t) = 1$ if the t-th observation falls in quarter i and $D_i(t) = 0$ otherwise. A stochastic model is

$$y(t) = y(t - 4) + \varepsilon(t). \tag{10.30}$$

This is a non-stationary AR(4) process with polynomial $\alpha(z) = 1 - z^{-4}$ that has four roots on the unit circle. A seasonal ARIMA model is of the form $\alpha(z^4)(1 - z^{-4})^d\, y = \beta(z^4)\varepsilon$, and mixtures of ARIMA and seasonal ARIMA models are also possible. A model with latent seasonal component is

$$S(t) = S(t - 4) + \varepsilon_1(t), \qquad y(t) = S(t) + \varepsilon_2(t). \tag{10.31}$$

The seasonal component $S(t)$ of this model can be estimated by the Kalman filter. The seasonal component can also be estimated by smoothing, for example

$$S(t) = \frac{1}{4}(y(t) + y(t - 1) + y(t - 2) + y(t - 3)). \tag{10.32}$$

If the time series contains trends and seasonals, then an additive model for this is given by

$$y(t) = T(t) + S(t) + R(t), \tag{10.33}$$

where T denotes the trend component, S the seasonal and R a stationary process. If trend and seasonal are proportional to the level of the series this can be expressed by the multiplicative model $y(t) = T(t)S(t)R(t)$, which gives a model of the form (10.33) by taking logarithms. If $\hat{T}(t)$ and $\hat{S}(t)$ are estimates of the trend and seasonal, then $\hat{R}(t) = y(t) - \hat{T}(t) - \hat{S}(t)$ is called the detrended and deseasonalized series. In many cases the series $\hat{R}$ is related to the original data y by means of a linear filter. The effect of this filter can be analysed in the frequency domain, where the data transformation is described in terms of spectra.

Proposition 10.4.1 *Let y be a stationary process with spectrum S, and consider the process $\overline{y}(t) = \sum_{k=-\infty}^{\infty} \beta_k y(t - k)$ with $\sum_{k=-\infty}^{\infty} \beta_k^2 < \infty$. Then the spectrum $\overline{S}$ of $\overline{y}$ is given by $\overline{S}(e^{i\omega}) = |\beta(e^{i\omega})|^2 S(e^{i\omega})$.*

Proof   Straightforward computation gives:

$$\overline{S}(e^{i\omega}) = \frac{1}{2\pi} \sum_{t=-\infty}^{\infty} e^{-it\omega} \sum_{j=-\infty}^{\infty} \sum_{k=-\infty}^{\infty} \beta_j \beta_k E\{y(t-j)y(t-k)\} =$$

$$= \frac{1}{2\pi} \sum_{j=-\infty}^{\infty} \beta_j e^{-ij\omega} \sum_{k=-\infty}^{\infty} \beta_k e^{ik\omega} \sum_{t=-\infty}^{\infty} R(t-j+k)e^{-i(t-j+k)\omega} =$$

$$= |\beta(e^{i\omega})|^2 S(e^{i\omega}). \qquad \square$$

So the effect of a filter is to reduce the importance of certain frequencies and to increase that of others. This is very useful in engineering applications, for example in communication where the low frequency signal is enhanced and the high frequency noise is suppressed. In economics this is of use to remove trends and seasonal effects. For example, if the observed time series is integrated as in (10.24) then the trend is removed by the transformation $\Delta y(t) = y(t) - y(t-1)$. This corresponds to the filter $c = 1 - e^{-i\omega}$, and the resulting spectrum is $S_{\Delta y}(e^{i\omega}) = 2(1 - \cos\omega)S(e^{i\omega})$. For frequency $\omega = 0$ the filter has value zero, so that the long term component is removed. In a similar way, the stochastic seasonal (10.30) can be removed by the transformation $\Delta_4 y(t) = y(t) - y(t-4)$, so that $S_{\Delta_4 y}(e^{i\omega}) = 2(1 - \cos(4\omega))S(e^{i\omega})$. The seasonal smoother (10.32) has filter $\beta(z) = 1 + z^{-1} + z^{-2} + z^{-3} = (1 - z^{-4})/(1 - z^{-1})$, so this corresponds to the spectral transformation $\frac{1}{16}\frac{1-\cos(4\omega)}{1-\cos(\omega)}$. As a comment on the difference between the filters corresponding to (10.30) and (10.32): the filter (10.30) models the short run fluctuations (high frequencies) and (10.32) the long run trend (low frequencies). The filter (10.30) is therefore called high pass, and (10.32) is called low pass.

In practice, the modelling of time series often proceeds in two steps. First the data are filtered to obtain stationarity, and then a model is estimated for the filtered data. As an example, suppose that an ARMAX model (10.20) is estimated for filtered input-output data $y_*(t) = \sum_k f_k y(t-k)$ and $u_*(t) = \sum_k f_k u(t-k)$. According to Theorem 10.2.6 it follows that, in terms of the original input series $u(t)$ and output series $y(t)$, the maximum likelihood estimator is given by

$$\int_{-\pi}^{\pi} \left| \frac{y(e^{i\omega})}{u(e^{i\omega})} - \frac{\beta(e^{i\omega})}{\alpha(e^{i\omega})} \right|^2 \cdot \left| \frac{\alpha(e^{i\omega})}{\gamma(e^{i\omega})} \right|^2 |f(e^{i\omega})|^2 S_{uu}(e^{i\omega})d\omega,$$

because $S_{u_*u_*}(e^{i\omega}) = |f(e^{i\omega})|^2 S_{uu}(e^{i\omega})$ according to Proposition 10.4.1. That is, prefiltering the data can be seen as a method to assign weights to the different frequencies in the identification criterion.

Trends and seasonals can also be seen as specific examples of time-varying parameters. For example, the models (10.23) and (10.29) describe time variations of the mean level of the series. For regression models, for example the ARX model (9.13), possible parameter variations can be analysed by recursive least squares. Let the regression model be written in the form $y(t) = x^T(t)\beta + \varepsilon(t)$ and let $\hat{\beta}(t)$ be the least squares estimate of β based on the observations $\{y(s), x(s), s \leq t\}$.

Theorem 10.4.2 *The recursive least squares estimates* $\hat{\beta}(t)$ *satisfy*

$$\hat{\beta}(t) = \hat{\beta}(t-1) + K(t)(y(t) - x^T(t)\beta(t-1)),$$

$$K(t) = P(t)x(t)/(1 + x^T(t)P(t)x(t)),$$

$$P(t+1) = P(t) - P(t)x(t)x^T(t)P(t)/(1 + x^T(t)P(t)x(t)),$$

where $P(t+1) = \text{var}(\hat{\beta}(t))$. *If* β *has* k *components, then the starting values are given by* $\hat{\beta}_k = (X_k^T X_k)^{-1} X_k^T y_k$ *and* $P(k+1) = (X_k^T X_k)^{-1}$, *where* y_k *and* X_k *contain the observations for* $t = 1, \cdots, k$.

Proof   The regression model can be written in state space form, with constant state vector $\beta(t+1) = \beta(t)$ and with output equation $y(t) = x^T(t)\beta(t) + \sigma\eta(t)$, where $\eta(t) = \varepsilon(t)/\sigma$ is standard white noise. In terms of the Kalman filter model (7.8), (7.9) the parameters are given by $A = I, B = 0, F = 0, D = 0, G = 0$, and with $C = x^T(t)$ known but time-varying. The Kalman filter equations of Theorem 7.3.1 also apply for non-stationary and time-varying systems. The given expressions then follow by the result in Chap. 9 (for the starting values), dividing $P(t)$ in these formulas by σ^2, and noting that $\hat{\beta}(t) = E(\beta|y(s), x(s), s \leq t)$ is $\hat{r}(t+1)$ in the notation of Theorem 7.3.1.   □

If the coefficients are constant, then the variance of the recursive residuals $\omega(t) = y(t) - x^T(t)\hat{\beta}(t-1)$ follows from Proposition 7.3.3. Therefore, $\omega^*(t) = \omega(t)/(1 + x^T(t)P(t)x(t))$ should be a white noise series. Several tests on parameter constancy have been developed that are based on the series $\omega^*(t)$. If the parameters turn out to be time varying, a possible model is the random walk $\beta(t+1) = \beta(t) + \eta(t)$, with η white noise. For given values of the regression variance $\sigma^2 = E(\varepsilon^2(t))$ and parameter variance $V = \text{var}(\eta(t))$, the parameters $\beta(t)$ can then be estimated recursively by the Kalman filter (Theorem 7.3.1) now with $F = V^{\frac{1}{2}}$. The parameters σ^2 and V can be estimated by maximum likelihood. An alternative way to deal with time-varying parameters is to apply weighted least squares, for example, with the criterion $\sum_{t=1}^N \alpha^{-t}\varepsilon^2(t)$ with $0 < \alpha < 1$. This assigns more weight to recent observations.

Time-varying parameters are one example of nonlinearity in observed time series. One of the possible causes is local linearization as shown in (1.21). As shown above, the estimates react more quickly to variations by applying weighted least squares. This can be generalized to local regression methods, where only the more recent observations are used in estimation. An alternative is to incorporate explicit nonlinear terms in the model.

That is, an ARMAX model like (10.20) is linear in the observed inputs and outputs and in the disturbances ε. Nonlinear models are of the type

$$y(t) = f(y(s - 1), u(s), \varepsilon(s), s \leq t) \qquad (10.34)$$

Higher order Taylor expansions of f may be useful, but in general the resulting number of parameters is too large for practical purposes. An alternative is to expand f in other basis functions that provide a more parsimonious description of the involved nonlinearities. Examples are neural networks and wavelets. Such models can be estimated, for instance, by nonlinear least squares methods. Theoretical knowledge on the general shape of the function f in (10.34) may be helpful in choosing an appropriate basis.

Further Developments

11

We close this book with a few sections that provide glimpses of further developments in the area of systems and control theory. In all sections we shall give a pointer to further literature on the subject.

11.1 Continuous Time Systems

In this book we have focussed our attention on systems in discrete time. The reason for this was that discrete time systems are in many ways easier to understand than continuous time systems, in particular this holds true for stochastic systems. We shall return to this issue in a later section.

Causal, linear, time-invariant input-output systems in continuous time can be modelled by a system of differential equations of the type

$$
\begin{aligned}
\dot{x}(t) &= Ax(t) + Bu(t), \\
y(t) &= Cx(t) + Du(t), \\
x(0) &= x_0.
\end{aligned}
\tag{11.1}
$$

Here, as in Chap. 2, A, B, C and D are matrices of appropriate sizes. Most of the theory of Chaps. 2, 3, and 4 holds in more or less the same way for continuous time systems. Obviously, there are differences as well: stability will hold in case A has all its eigenvalues in the open left half plane (instead of the open unit circle). Also, the Stein equations of

Chap. 4 have to be replaced by Lyapunov equations of the type

$$A^*P + PA = -C^*C,$$

$$AQ + QA^* = -BB^*.$$

(11.2)

Continuous time systems have traditionally been very important in applications in engineering, where models are usually built from first principles, that is, from the description of physical components in mathematical models that involve differential equations (like in mechanics and electronics). The theory is well explained in many standard textbooks in systems theory, see for instance [27, 37, 58, 76].

Other approaches to continuous time systems, using more the transfer function and input-output operator, rather then state space models have also been influential. See for instance the book [17].

In certain applications the state is not only subject to differential equations, but also to algebraic equations. Situations like that can be modelled by so-called descriptor systems, of the type

$$E\dot{x}(t) = Ax(t) + Bu(t),$$

$$y(t) = Cx(t) + Du(t),$$

$$x(0) = x_0.$$

Descriptor systems have been studied in detail in the literature, a good source is [43].

Time varying and periodic systems also occur in many applications. In particular periodicity has been studied in connection with technical applications that require a periodical behaviour. Such systems are usually modelled as in (11.1) where instead of fixed matrices A, B, C and D, these matrices are taken to be time varying or periodic.

In recent decades the view of systems has changed from an input-output view to a view using only external and internal variables. This point of view is particularly useful in certain applications, and we have tried to show some of this point of view in our first chapter when we discussed the system behaviour. Systems theory developed from this point of view is sometimes called the behavioural approach, see [62].

11.2   Optimal Control

Optimal control as outlined in Chaps. 5 and 8 has been one of the topics that most influenced the development of control theory. The program of missions to the moon and to the planets would not have been possible without substantial developments in optimal control theory. It is precisely in these applications that optimal control in state space terminology was so successful. Besides LQ-optimal control the topic of time-optimal control is well studied and well-described in the literature. In the time-optimal control problem the problem is to reach a given target state from a given initial state in the shortest possible time, under restrictions on the size of the input. See [30, 50] for good elementary treatments.

There are many different approaches to optimal control theory. One is the use of dynamic programming, as we have done here. Others make use of the Pontryagin optimality principle. See [1, 30, 50].

For optimal control of stochastic systems we refer to [7, 8].

As an example of results in this area, let us state here the main result of the infinite horizon LQ-optimal control for a continuous time system of the form

$$\dot{x}(t) = Ax(t) + Bu(t),$$
$$x(0) = x_0, \tag{11.3}$$

with cost function given by

$$J(x_0, u) = \int_0^\infty x(t)^* Qx(t) + u(t)^* Ru(t)\, dt. \tag{11.4}$$

As in the discrete time case we shall assume that (A, B) is stabilizable, Q is positive semidefinite and R is positive definite. The goal is to find the minimum of $J(x_0, u)$ over all stabilizing input functions $u(t)$, and to find the minimizing input function. As in the discrete time case, there is a matrix equation to be solved, in this case too it is called the (continuous) algebraic Riccati equation. The result is as follows.

Theorem 11.2.1 *Assume that (A, B) is stabilizable and that (A, Q) is detectable. Then the minimum of* (11.4) *subject to* (11.3) *is given by $x_0^* X x_0$, where X is the unique solution of the algebraic Riccati equation*

$$XBR^{-1}B^* X - XA - A^* X - Q = 0 \tag{11.5}$$

for which the closed loop matrix $A - BR^{-1}B^* X$ is asymptotically stable. In this case the minimizing input is given by the static state feedback $u(t) = -R^{-1}B^* Xx(t)$.

Note that here we fix the endpoint, that is, we fix $\lim_{t\to\infty} x(t)$ to be zero. Other possibilities also have been considered, both in the continuous time and in discrete time case. For instance, problems with indefinite cost (that is with Q and R possibly indefinite), problems with free endpoint or with endpoint constrained to be in a given subspace. See [75, 79] for the continuous time case and in the discrete time case [63].

11.3   Nonlinear Systems

Nonlinear systems are typically studied in the form of a system of nonlinear differential equations coupled to an output equation. In wide generality such systems can be described as follows:

$$\dot{x}(t) = f(x(t), u(t), t),$$

$$x(0) = x_0,$$

$$y(t) = g(x(t), u(t), t).$$

Stability of equilibrium solutions can then be discussed using methods from the theory of ordinary differential equations. A common approach to studying the systems would be to linearize around the equilibrium solutions, and for each equilibrium solution one arrives at a linear system.

In many examples the control variable u enters in a linear way, and the functions f and g are time invariant. That leads to systems of the form

$$\dot{x}(t) = f(x(t)) + h(x(t)) \cdot u(t),$$

$$x(0) = x_0,$$

$$y(t) = g(x(t)) + k(x(t)) \cdot u(t).$$

As a simple example, consider the pendulum of the following figure

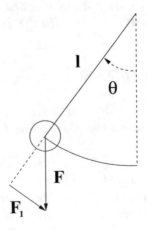

A pendulum of length l, mass m hangs on an axis. We suppose friction plays no role (that is why we call this the "mathematical" pendulum). Denote by $\theta(t)$ the deviation of the pendulum from the downward vertical. The differential equation for $\theta(t)$ that governs the motion of the pendulum under influence of gravity is given by

$$\theta''(t) = -\frac{g}{l}\sin(\theta(t)).$$

As initial conditions we take $\theta(0) = \theta_0, \theta'(0) = \theta_0'$. This has two equilibrium solutions, the stable equilibrium being the restposition in the downward position, the unstable one being the upright position. Now suppose that on the axis we can put a torque.

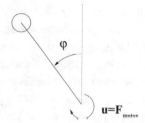

This torque is viewed as an input. In terms of $\varphi(t) = \pi - \theta(t)$ the differential equation becomes

$$\varphi''(t) = \frac{g}{l}\sin(\varphi(t)) + c \cdot u(t),$$

where c is a constant. The goal is to show that it is possible to find a control function that will bring the pendulum in upright position for small deviations of that equilibrium position.

Nonlinear systems theory has many important applications. In engineering we mention applications in robotics. But also recent activity in bio-medical sciences has lead to interesting applications of nonlinear systems.

There are many excellent books on nonlinear systems theory. We mention here [34, 35, 41, 57, 66, 76, 80].

An important issue in some control problems, both linear and nonlinear, is positivity. Several applications, notably those where the states are concentrations of substances, require state variables to be always nonnegative. This is a very difficult issue that is still under research. See for instance [6, 31].

Applications in engineering frequently also involve some form of energy considerations, e.g. energy conservation or dissipation of energy. This point of view has led to a new view on systems and control theory in the framework of Port-Hamiltonian systems [67, 68].

11.3.1 Applications in Life Sciences

It has long been recognized that the concept of feedback plays a role in many biochemical pathways in the cell. With the advent of better understanding of these pathways using mathematical modeling there is scope for applying nonlinear control theory to problems in biology. For instance, metabolical pathways in the cell can now be modelled by large scale systems of (relatively simple) nonlinear differential equations of the Michaelis-Menten type, coming from reaction kinetics. Understanding the system as a whole is then a formidable task, as there are really many equations involved, and it is at this point that systems theory may help by considering methods of reducing the model to a smaller one (model reduction), and still keep the most salient behaviour.

As an example, we refer to the mathematical model for glycolysis in Trypanosoma brucei, the organism responsible for sleeping sickness. A mathematical model for glycolysis in this organism was developed and studied with a view to use this as a basis for control. Steps to apply systems and control theory to further understand such models coming from biochemical pathways were taken in [56] (see also references give there), [61, 77, 81].

11.4   Infinite Dimensional Systems

Much of the theory described in the first five chapters of this book, i.e., the deterministic part, has a counterpart in continuous time infinite dimensional systems. For example, systems that require a physical description using a partial differential equation (like the equation for a vibrating string, or the heat equation), or systems with a delay in the time argument, can usually be modelled quite well in the form (11.1), where A is the generator of a C_0-semigroup on a Banach or Hilbert space, and B and C are linear (possibly also unbounded) operators acting between the input and state space, respectively, the state space and the output space.

Continuous time systems that are only observed at regular time intervals (sampled) and controlled at the same time intervals, after which the control is kept constant until the next sampling time, are commonly known as sampled data systems. Such systems can be modelled fruitfully in terms of a discrete time system with an infinite dimensional state space.

The theory of infinite dimensional systems requires a solid background in functional analysis. An excellent place to start when learning this subject is the book [14]. This book deals with both a state-space approach and a frequency domain approach (i.e., using the transfer function as the main tool for the study of the system). We also mention the two books [4, 5]. A different approach is to study the systems entirely from the point of view of partial differential equations. For this point of view, see e.g., [46].

As a sample of the kind of systems that is studied in the theory of infinite dimensional systems consider the following delay system:

$$\dot{x}(t) = ax(t) + bx(t-1) + \int_{t-1}^{t} u(\tau)\, d\tau.$$

The initial condition for such an equation needs to be a function on the interval $[-1, 0]$, and hence equations of this type have a state space that must be some function space over that interval. That may be the Banach space of continuous functions on $[-1, 0]$, or the Hilbert space of L_2 functions on that interval, or some other Banach space. For details, see [14, 23].

For infinite dimensional systems too the framework of Port-Hamiltonian systems plays an important role in new developments, in particular with a view to applications in systems governed by partial differential equations. We refer to [36] for a mathematically oriented introduction into the subject with many applications mentioned there as well.

11.5 Robust and Adaptive Control

Uncertainty in systems has been described in this book largely in terms of stochastic additive uncertainties. However, in practice issues like unmodelled dynamics often lead to other types of uncertainties. For instance, one can think of the given model as an approximation of the "true" model in a certain neighbourhood of the true model. Robust control methods then strive to design a controller that not only will stabilize the given model, but also all models in a given neighbourhood. Also other design criteria, besides stabilizing the system, are considered.

This all depends heavily on the way the model is given, and in modern H^∞ control the model is usually considered to be given as a transfer function. This naturally leads to the fact that robust control theory is a theory in the frequency domain to start with. See, e.g., [17, 54]. However, very nice results have been obtained in the state space framework as well. See, e.g., [18, 22, 83], the later chapters in [27] and [3], Chapters 19 and 20. Most of this is developed for continuous time systems. For discrete time systems, see also [26, 33, 78].

In a sense, also adaptive control does the same: its goal is to stabilize a large class of systems with a single controller. The main idea is that this may be possible using a nonlinear controller that adapts to the unknown parameters in the system. For instance, in this way linear first order systems of the type $\dot{y} = ay + bu$, with $b \neq 0$ but otherwise unknown, can all be stabilized with a single control algorithm. For more developments in this direction see, e.g., [53].

As a sample of results in the area of H^∞ control, consider the following problem: given is a continuous time system with two inputs (w and u) and two outputs (y and z) as in the following figure.

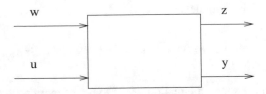

We consider u as usual as the input that we can control, w as the disturbances, y as the measurements we can take, and z as the output that we wish to control. In this section we study the full information case, that is we take $y = x$. The system is then given by the equations

$$\dot{x}(t) = Ax(t) + B_1 w(t) + B_2 u(t),$$

$$z(t) = Cx(t) + Du(t),$$

$$y(t) = x(t).$$

The goal is to make the influence of w on z small in an appropriate measure, which we shall make more precise below.

We consider state feedback $u(t) = Kx(t)$, where K is a constant matrix.

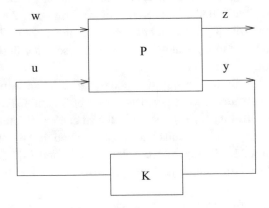

Hence the closed loop system is determined by the following equations:

$$\dot{x}(t) = (A + B_2 K)x(t) + B_1 w(t),$$

$$z(t) = (C + DK)x(t).$$

Let $G_K(s)$ denote the transfer function from w to z, that is, $G_K(s) = (C + DK)(sI - (A + B_2K))^{-1}B_1$. Then we want to find K such that the following two conditions hold: 1. for some pre-specified tolerance level γ we have

$$\|G_K\|_\infty := \max_{s \in i\mathbb{R}} \|G_K(s)\| < \gamma$$

and 2. in addition K is a stabilizing feedback, that is,

$$A + B_2K \text{ is stable.}$$

We shall make the following assumptions:

1. the pair (C, A) is observable,
2. the pairs (A, B_1) and (A, B_2) are stabilizable,
3. $D^T C = 0$ and $D^T D = I$.

Note that the first two assumptions are not so unnatural, but that the third one may seem a little strange. However, it can be proven that this can always be achieved by applying a state feedback at the start, as long as we assume that D has full column rank. Indeed, in that case $D^T D$ is invertible, and we can assume without loss of generality that $D^T D = I$, as this only entails a change of bases in the input space and in the output space. After that, consider applying a feedback with feedback matrix $K = -D^T C$, so that C is replace by $C + DK$. Then $D^T(C + DK) = D^T C + K = 0$. Under these assumptions we have the following theorem.

Theorem 11.5.1 *There exists a matrix K such that $A + B_2K$ is stable and $\|G_K\|_\infty < \gamma$ if and only if there exists a positive definite matrix X_∞ for which the following two conditions are met:*

1. *X_∞ satisfies the algebraic Riccati equation*

$$X(\tfrac{1}{\gamma^2}B_1 B_1^T - B_2 B_2^T)X + XA + A^T X + C^T C = 0,$$

2. *$A + (\tfrac{1}{\gamma^2}B_1 B_1^T - B_2 B_2^T)X_\infty$ is stable.*

In that case one such state feedback is given by $K = -B_2^T X_\infty$.

It may be observed that if $\gamma \to \infty$ then X_∞, considered as a function of γ will go to the solution of the LQ-optimal control problem.

11.6 Stochastic Systems

Most applications of stochastic systems in engineering and economics employ discrete time models. This is because the data are often recorded at discrete time instants and because stochastic processes in discrete time are somewhat simpler to analyze and implement than processes that evolve in continuous time. Stochastic systems in discrete time and their applications in system estimation and control are discussed, for instance, in [12, 16, 42].

Finite dimensional continuous time stochastic systems are described by a set of stochastic differential equations of the form

$$dx(t) = Ax(t)dt + Bu(t)dt + Ed\varepsilon$$
$$y(t) = Cx(t) + Du(t) + F\varepsilon(t)$$

where ε is a continuous time white noise process. This is the continuous time analogue of the stochastic input-output system (6.37) described in Sect. 6.6, obtained by adding a noise process to the deterministic continuous time system (11.1) of Sect. 11.1. For given input trajectory $u(t)$, the solution processes for $x(t)$ and $y(t)$ of the above system of equations are defined in terms of stochastic integrals.

Systems of this type are used in (continuous time) stochastic control. Another area of application is mathematical finance, where price movements of financial assets and derivatives are modelled in this way. For instance, the Black-Scholes formula for option prices is based on the assumption that stock prices follow a Brownian motion, see e.g. [32]. The analysis and solution of such continuous time stochastic systems is based on the theory of stochastic differential equations, see e.g. [2] and [15] and for applications in mathematical finance [38, 73].

11.7 Networked Systems

Modern control systems are frequently highly structured. In many applications the overall system consists of a hierarchical structure, with a central coordinating system, which controls subsystems, which in turn, may act as coordinator for subsystems at a lower level. Communication restrictions may exist between the systems at the lower level and the systems at the higher level. In particular, communication may use a network such as internet or a mobile communication network, on which communication is uncertain and subject to delays.

Examples of this are to be found, e.g., in communication between cars on highways to form platoons, the development of self-driving cars, and control of systems of which the architecture may change, via e.g. telecommunication networks.

There are several sources on this topic, of which we mention [39, 51, 70–72] and the references given there.

11.8 Hybrid Systems

Hybrid systems are systems where continuous or discrete time systems may switch between different regimes, triggered by a discrete event. Because of the latter point, such systems differ from time-varying systems, as in that case the system matrices may be time-varying, but they are explicitly known as functions of time. In a hybrid system that is not the case: the changes in the system dynamics may occur for instance when the state reaches a certain boundary, which triggers the dynamics to change drastically. Such systems are not easy to analyse and control, but play an important role in many applications. A good introduction to the subject is [69], a thorough discussion of realization theory for such systems may be found in [60].

11.9 System Identification

System identification covers a very broad area, because modelling dynamical phenomena from observed data is applied in numerous fields that are as diverse as e.g., astronomy, micro biology, psychology, management. Therefore it is not well possible to provide a brief overview of even just the main developments. However, one common characteristic is that the development of more advanced methods goes hand in hand with the tremendous growth in computing power. This allows the modelling of very large data sets (for instance in biology and in finance and marketing) and the development of more advanced (non-linear) models.

We briefly mention some issues of particular interest in two application areas, engineering and economics. In engineering, nonlinear models are employed to incorporate nonlinear response, e.g., due to saturation effects. For controlled systems the issue of closed loop identification is of importance, as the applied regulator affects the observed system dynamics. Further, apart from the more conventional input-output based estimation methods, one sometimes also uses state space models in (so-called) subspace identification techniques. For more background on system identification in engineering we refer to [48] and [74].

In economics, the availability of large data sets in areas like finance and marketing allows the estimation of more and more elaborate models to describe the movements of economic variables like prices and sales. Recent developments include the modelling of trending patters and changes in volatility and risk. More background on these and other issues in modern business and economics can be found, e.g., in [11, 29, 59] and [82].

Bibliography

1. B.D.O. Anderson, J.B. Moore, *Optimal Control: Linear Quadratic Problems* (Prentice Hall, Engelwood Cliffs, NJ, 1990)
2. L. Arnold, *Stochastic Differential Equations: Theory and Applications* (Wiley, New York, 1974)
3. H. Bart, I. Gohberg, M.A. Kaashoek, A.C.M. Ran, *A State Space Approach to Canonical Factorization with Applications.* Operator Theory Advances and Applications, Vol. 200 (Birkhäuser, Basel, 2010)
4. A. Bensoussan, G. Da Prato, M.C. Delfour, S.K. Mitter, *Representation and Control of Infinite-Dimensional Systems*, Vol. 1 (Birkhäuser Boston, Boston, MA, 1992)
5. A. Bensoussan, G. Da Prato, M.C. Delfour, S.K. Mitter, *Representation and Control of Infinite-Dimensional Systems*, Vol. 2 (Birkhäuser Boston, Boston, MA, 1993)
6. A. Berman, M. Neumann, R.J. Stern, *Nonnegative Matrices in Dynamical Systems* (Wiley, New York, 1989)
7. D.P. Bertsekas, *Dynamic Programming. Deterministic and Stochastic Models* (Prentice Hall, Englewood Cliffs, NJ, 1987)
8. D.P. Bertsekas, S.E. Shreve, *Stochastic Optimal Control. The Discrete Time Case* (Academic Press, New York, London, 1978)
9. P.P.J. van den Bosch, A.C. van der Klauw, *Modeling, Identification and Simulation of Dynamical Systems* (CRC Press, Boca Raton, FL, 1994)
10. P.J. Brockwell, R.A. Davis, *Time Series: Theory and Methods* (Springer, New York, 1987)
11. C. Brooks, *Introductory Econometrics for Finance* (Cambridge University Press, Cambridge, 2002)
12. P.E. Caines, *Linear Stochastic Systems* (Wiley, New York, 1988)
13. H. Cramér, *Mathematical Methods of Statistics.* Princeton Mathematical Series, vol. 9 (Princeton University Press, Princeton, NJ, 1946)
14. R.F. Curtain, H.J. Zwart, *An Introduction to Infinite-Dimensional Linear Systems Theory* (Springer, New York, 1995)
15. M.H.A. Davis, *Linear Estimation and Stochastic Control* (Chapman and Hall, London, 1977)
16. M.H.A. Davis, R.B. Vinter, *Stochastic Modelling and Control* (Chapman and Hall, London, 1985)
17. J.C. Doyle, B.A. Francis, A.R. Tannenbaum, *Feedback Control Theory* (Macmillan Publishing, New York, 1992)
18. J.C. Doyle, K. Glover, P.P. Khargonekar, B.A. Francis, State-space solutions to standard H_2 and H_∞ control problems. IEEE Trans. Automat. Control **34**(8), 831–847 (1989)
19. I. Gohberg, P. Lancaster, L. Rodman, *Matrix Polynomials* (Academic Press, New York, 1982)

20. G.H. Golub, C.F. Van Loan, *Matrix Computations*, 3rd edn. Johns Hopkins Studies in the Mathematical Sciences (Johns Hopkins University Press, Baltimore, MD, 1996)
21. G. Goodwin, R.L. Payne, *Dynamic System Identification. Experiment Design and Data Analysis* (Academic Press, New York, London, 1977)
22. M. Green, K. Glover, D. Limebeer, J.C. Doyle, A J-spectral factorization approach to H_∞ control. SIAM J. Control Optim. **28**(6), 1350–1371 (1990)
23. J.K. Hale, S.M. Verduyn Lunel, *Introduction to Functional Differential Equations* (Springer, New York, 1993)
24. J.D. Hamilton, *Time Series Analysis* (Princeton University Press, Princeton, NJ, 1994)
25. E.J. Hannan, M. Deistler, *The Statistical Theory of Linear Systems* (Wiley, New York, 1988)
26. B. Hassibi, A.H. Sayed, T. Kailath, *Indefinite-Quadratic Estimation and Control, A Unified Approach to H^2 and H^∞ Theories* (SIAM, Philadelphia, PA, 1999)
27. M. Hautus, H.L. Trentelman, A.A. Stoorvogel, *Control Theory for Linear Systems* (Springer, New York, 2001)
28. C. Heij, *Deterministic Identification of Dynamical Systems*. LNCIS, 127 (Springer, Berlin, 1989)
29. C. Heij, P.M. De Boer, P.H. Franses, T. Kloek, H.K. Van Dijk, *Econometric Methods with Applications in Business and Economics* (Oxford University Press, Oxford, 2004)
30. L.M. Hocking, *Optimal Control, An Introduction to the Theory with Applications* (Oxford Univ. Press, Oxford, 1991)
31. J.M. van den Hof, Realization of positive linear systems. Linear Algebra Appl. **256**, 287–308 (1997)
32. J.C. Hull, *Options, Futures, and Other Derivative Securities* (Prentice Hall, Engelwood Cliffs, 2002)
33. V. Ionescu, C. Oarǎ, M. Weiss, *Generalized Riccati Theory and Robust Control, A Popov Function Apporach* (Wiley, Chichester, 1999)
34. A. Isidori, *Nonlinear Control Systems* (Springer, Berlin, 1995)
35. A. Isidori, *Nonlinear Control Systems, II* (Springer, London, 1999)
36. B. Jacob, H. Zwart *Linear Port-Hamiltonian Systems on Infinite Dimensional Spaces*. Operator Theory Advances and Applications, Vol. 223 (Birkhäuser, Basel, 2012)
37. T. Kailath, *Linear Systems* (Prentice Hall, Englewood Cliffs, NJ, 1980)
38. I. Karatzas, S.E. Shreve, *Brownian Motion and Stochastic Calculus*, 2nd edn. (Springer, New York, 1991)
39. P. Kempker, *Coordination Control of Linear Systems*. Ph.D. thesis, Vrije Universiteit Amsterdam, 2012
40. M. Kendall, A. Stuart, *The Sdvanced Theory of Statistics. Vol. 2: Inference and Relationship*, 2nd edn. (Hafner Publishing, New York, 1967)
41. H.K. Khalil, *Nonlinear Systems*, 3rd edn. (Prentice Hall, Engelwood Cliffs, NJ, 2002)
42. P.R. Kumar, P.P. Varaiya, *Stochastic Systems: Estimation, Identification and Adaptive Control* (Prentice Hall, Englewood Cliffs, NJ, 1986)
43. P. Kunkel, V. Mehrmann, *Differential-Algebraic Equations – Analysis and Numerical Solution* (EMS Publishing House, Zürich, 2006)
44. H. Kwakernaak, R. Sivan, *Linear Optimal Control Systems* (Wiley, New York, London, Sydney, 1972)
45. P. Lancaster, L. Rodman, *Algebraic Riccati Equations* (Oxford University Press, Oxford, 1995)
46. I. Lasiecka, *Mathematical Control Theory of Coupled PDEs* (SIAM, Philadelphia, PA, 2002)
47. L. Ljung, System identification: design variables and the design objective, in *Modelling, Robustness and Sensitivity Reduction in Control Systems (Groningen, 1986)*. NATO Adv. Sci. Inst. Ser. F Comput. Systems Sci., vol. 34 (Springer, Berlin, 1987), pp. 251–270
48. L. Ljung, *System Identification - Theory For the User* (Prentice Hall, Engelwood Cliffs, 1999)

49. L. Ljung, T. Söderström, *Theory and Practice of Recursive Identification* (MIT Press, Cambridge, MA, London, 1983)
50. A. Locatelli, *Optimal Control* (Birkhäuser, Basel, 2001)
51. J. Lunze, *Networked Control of Multi-Agent Systems: Consensus and Synchronisation, Communication Structure Design, Self-organisation in Networked Systems, Event-Triggered Control* (Bookmundo, 2019)
52. H. Lütkepohl, *Introduction to Multiple Time Series Analysis*, 2nd edn. (Springer, Berlin, 1993)
53. I. Mareels, J.W. Polderman, *Adaptive Systems. An Introduction* (Birkhäuser Boston, Boston, MA, 1996)
54. D.C. McFarlane, K. Glover, *Robust Controller Design Using Normalized Coprime Factor Plant Descriptions*. LNCIS 138 (Springer, Berlin, 1990)
55. M. Moonen, B. de Moor, J. Ramos, S. Tan, A subspace identification algorithm for descriptor systems. Syst. Control Lett. **19**, 47–52 (1992)
56. J. Němcová, M. Petreczky, J.H. van Schuppen, Towards a system theory of rational systems. Oper. Theory Adv. Appl. **271**, 327–359 (2018)
57. H. Nijmeijer, A.J. van der Schaft, *Nonlinear Dynamical Control Systems* (Springer, New York, 1990)
58. G.J. Olsder, J.W. van der Woude, *Mathemtical Systems Theory* (Delft University Press, Delft, 2005)
59. K. Patterson, *An Introduction to Applied Econometrics* (Palgrave, London, 2000)
60. M. Petreczky, *Realization Theory of Hybrid Systems*. Ph.D. thesis, Vrije Universiteit Amsterdam, 2006
61. R. Planqué, J. Hulshof, B. Teusink, J.C. Hendriks, F.J. Bruggeman, Maintaining maximal metabolic rates by gene expression control. PLoS Comput. Biol. **14**(9), e1006412 (2018)
62. J.W. Polderman, J.C. Willems, *Introduction to Mathematical Systems Theory. A Behavioral Approach* (Springer, New York, 1998)
63. A.C.M. Ran, H.L. Trentelman, Linear quadratic problems with indefinite cost for discrete time systems. SIAM J. Matrix Anal. Appl. **14**(3), 776–797 (1993)
64. S. Ross, *A First Course in Probability* (Macmillan Publishing, New York, 1984)
65. W. Rudin, *Principles of Mathematical Analysis*, 3rd edn. (McGraw-Hill, New York, 1976)
66. S. Sastry, *Nonlinear Systems - Analysis, Stability and Control* (Springer, Berlin, 1999)
67. A.J. van der Schaft, *Port-Hamiltonian Systems: An Introductory Survey*. In International Congress of Mathematicians, Vol. III (EMS Publ., Zürich, 2006), pp. 1339–1365
68. A.J. van der Schaft, *L_2-Gain and Passivity Techniques in Nonlinear Control*, 3rd edn. Communications and Control Engineering Series (Springer, Cham, 2017)
69. A.J. van der Schaft, J.M. Schumacher, *An Introduction to Hybrid Dynamical Systems*. Lecture Notes in Control and Information Sciences, vol. 251 (Springer, London, 2000)
70. J.H. van Schuppen, O. Boutin, P.L. Kempker, J. Komenda, T. Masopust, N. Pambakian, A.C.M. Ran, Control of distributed systems: tutorial and overview. Eur. J. Control **17**, 579–602 (2011)
71. J.H. van Schuppen, T. Villa (eds.), *Coordination Control of Distributed Systems* (Springer, New York, 2015)
72. P. Shah, P.A. Parrilo, A partial order approach to decentralized control, in *Proc. 47th IEEE Conference on Decision and Control* (CDC.2008), Cancun, Mexico (2008)
73. S.E. Shreve, *Stochastic Calculus for Finance. II. Continuous-Time Models* (Springer, New York, 2004)
74. T. Söderström, *Discrete-Time Stochastic Systems. Estimation and Control*, 2nd edn. (Springer, London, 2002)
75. J.M. Soethoudt, H.L. Trentelman, The regular indefinite linear-quadratic problem with linear endpoint constraints. Syst. Control Lett. **12**(1), 23–31 (1989)

76. E.D. Sontag, *Mathematical Control Theory, Deterministic Finite Dimensional Systems* (Springer, New York, 1998)
77. E.D. Sontag, Structure and stability of certain chemical networks and applications to the knietic proofreading model of T-cell receptor signal transduction. IEEE Trans. Automatic Control **46**, 1028–1047 (2001)
78. A.A. Stoorvogel, *The H-Infinity Control Problem: A State Space Approach* (Prentice-Hall, Englewood Cliffs, 1992)
79. H.L. Trentelman, The regular free-endpoint linear quadratic problem with indefinite cost. SIAM J. Control Optim. **27**(1), 27–42 (1989)
80. M. Vidyasagar, *Nonlinear Systems Analysis*. Reprint of the second (1993) edition (SIAM, Philadelphia, PA, 2002)
81. H.V. Westerhoff, M. Verma, M. Nardelli, M. Adamczyk, K. van Eunen, E. Simeonidis, B.M. Bakker, Systems biochemistry in practice: experimenting with modelling and understanding, with regulation and control. Biochem. Soc. Trans. **38**, 1189–1196 (2010)
82. J.M. Wooldridge, *Econometric Analysis of Cross Section and Panel Data* (Cambridge, MIT, 2002)
83. K.H. Zhou, *Essentials of Robust Control* (Prentice Hall, Engelwood Cliffs, NJ, 1998)

Index

A
Aliasing, 162
Anticausal filter, 88
ARMA model, 86
ARMAX, 82
AR-representation, 85
Asymptotic stability, 49

B
Backward shift, 23
Bang-bang control, 72
BIBO stable, 54

C
Causal filter, 88
Causal system, 5, 13
Certainty equivalence, 122
Closed loop system, 57
Controllability Grammian, 52
Controllable, 27
Controllable eigenvalue, 34
Cost-to-go functions, 69

D
Dead-beat controller, 58
Detectable, 61
Dilation, 34
Dynamical system, 5
Dynamic compensator, 60
Dynamic programming, 68

E
Ergodic process, 83
Estimator, BLUE, 139
Estimator, consistent, 138
Estimator, efficient, 138
Estimator, linear, 139
Estimator, unbiased, 138
Exponentially bounded, 15
Externally stable, 54
External matrix, 17

F
Feedthrough matrix, 17
Filter, 86
Filtering, 101
Finite support sequence, 23
Frequency domain, 16

G
Gain, 16

H
Harmonic process, 85
Hautus test, 34

I
Impulse, 14
Impulse response, 14
Infinite horizon LQ problem, 75

© The Author(s), under exclusive license to Springer Nature Switzerland AG 2021
C. Heij et al., *Introduction to Mathematical Systems Theory*,
https://doi.org/10.1007/978-3-030-59654-5

Printed in the United States
by Baker & Taylor Publisher Services